Detectors for particle radiation

KONRAD KLEINKNECHT *University of Mainz*

Detectors for particle radiation

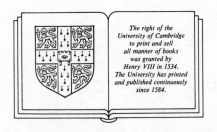

The right of the
University of Cambridge
to print and sell
all manner of books
was granted by
Henry VIII in 1534.
The University has printed
and published continuously
since 1584.

CAMBRIDGE UNIVERSITY PRESS

Cambridge

London   New York   New Rochelle

Melbourne   Sydney

Published by the Press Syndicate of the University of Cambridge
The Pitt Building, Trumpington Street, Cambridge CB2 1RP
32 East 57th Street, New York, NY 10022, USA
10 Stamford Road, Oakleigh, Melbourne 3166, Australia

© Cambridge University Press 1986

First published 1986

Printed in Great Britain at the University Press, Cambridge

British Library cataloguing in publication data

Kleinknecht, Konrad
Detectors for particle radiation.
1. Nuclear counters
I. Title
539.7'7   QC787.C6

Library of Congress cataloguing in publication data

Kleinknecht, K. (Konrad), 1940–
Detectors for particle radiation.
Bibliography: p.
Includes index.
1. Particles (Nuclear physics) – Instruments.
2. Nuclear counters. I. Title.
QC786.K57 1986   539.7'2'0287   85-17502

ISBN 0 521 30424 5

MP

CONTENTS

....

PREFACE

....

Progress in a branch of experimental physics is always closely linked with improved methods of measurement in this field. In searching for the elementary constituents of matter and for the forces between them, physicists use particle accelerators and detectors as tools to obtain the reaction products produced in collisions between elementary particles. These reaction products are either massive particles or the quanta of electromagnetic radiation.

Accelerators correspond to the microscope of the scientist, except that the probe used is not visible light but a charged particle like the electron, the proton or a heavy ion. Because of the duality between particles and waves, light and charged particles can both be used as probes. With increasing energy of the particles, their wavelength decreases, and so does the size of objects which can be resolved by the particle microscope. In this way, the search for ever-smaller objects required the construction of larger and larger accelerators. In addition, technological progress in accelerator construction was made, including the invention of stochastic cooling of antiproton beams and the development of superconducting pulsed dipole magnets.

Similarly, the methods for detecting particles or radiation have developed rapidly in all fields of application: detectors are needed in particle physics experiments, nuclear physics experiments, nuclear medicine, cosmic-ray measurements, and geological exploration. While a rich literature exists on accelerator principles and design, the same is not true for the subject of particle detectors. In particular, an introductory description of the more recent developments is missing.

The content of this book is based on courses of lectures given at Dortmund University since 1974. In addition, I had the opportunity to work on the subject during two summer schools in 1980: one at Zakopane

organized by A. Bialas, the other one at St. Croix (USVI) at the invitation of
T. Ferbel. I thank the organizers for their contribution to the genesis of this
book, my colleagues D. Wegener (Dortmund), A. Wagner (Heidelberg),
W. Blum (München) and J. May (Geneva) for discussions and useful hints;
Mrs. E. Lorenz for typing the manuscript; and Mrs. H. Bußmann and Mr. J.
Huhn for drawing and reproducing the diagrams.

Dortmund, December 1984   K. Kleinknecht

1

Physics foundations

1.1 Range of application for radiation detectors

1.1.1 Natural sources of radiation

Ionizing radiation originates either from our natural environment or from artificial sources. In both cases, the primary radiation consists of massive charged particles or of massless neutral quanta, i.e. photons or neutrinos.

If we observe radiation from our environment, two main origins are in evidence: one is the cosmic and solar radiation which comes from space and impinges on the outer part of the earth's atmosphere. It consists mainly of protons, light nuclei and electrons. By interacting with the terrestrial atmosphere, it produces secondary particles, including short-lived π mesons and muons. The study of the composition and energy distribution of primary cosmic radiation is best achieved above the atmosphere. Balloon flights, satellites and Space Shuttle flights are used for this research.

The most intensive source of radiation near to our planet is, of course, the sun. Apart from the visible light from its surface and the neutrinos from the cyclic nuclear fusion reactions in the core, it also emits massive particles; these are mainly electrons and protons ejected during eruptions and flares from the surface. When arriving at the earth, they cause zodiacal light and magnetic storms. This 'solar wind' of particles is also responsible for the radiation belts around the earth.

The other natural source of radiation, as discovered by Becquerel in 1896 in uranium ores, is natural α-, β-, or γ-radioactivity. α-Decay is a common phenomenon amongst heavy nuclei. Such a nucleus can emit an α-particle (or ^{4}He nucleus) which penetrates the Coulomb barrier of the heavy nucleus by tunnelling. It is mono-energetic with a kinetic energy in the range of 2–10 MeV. β-Decay is the transformation of a nucleus ^A_ZX with Z protons and $(A–Z)$ neutrons into another one, $^A_{z+1}$X, thereby emitting an electron e$^-$

and an antineutrino $\bar{v}$. The energy spectrum of the electrons is continuous, ranging from zero to the endpoint energy, given by the difference between the energies of the nuclear levels of the mother and daughter nuclei. Endpoint energies up to few MeV are observed. γ-Decay is the name for the decay of an excited nuclear state $^A_ZX^*$ into another state of the same nucleus ^A_ZX. In this process, mono-energetic hard γ-rays with energies in the MeV range are emitted.

Natural radioactivity is of crucial importance for the interior energy balance of the earth. The only radioactivity accessible for measurement is that in the crust of the earth, which is explored by drillings up to a depth of 5 km. Such measurements can give information about beds of certain minerals and are used in searches for uranium ores or petroleum.

Natural radioactivity can also be used for age determination of minerals in terrestrial rocks, meteorites and material from the moon. They fix the age of our planet at about 4.5×10^9 years. For organic matter, dating can be achieved by measuring the β-activity of the carbon isotope ^{14}C. In the atmosphere, the concentration of ^{14}C is kept in equilibrium by the continuous production through cosmic radiation and its decay with a half-life of 5730 years. If a plant or animal dies, the exchange of CO_2 with the atmosphere ceases and the concentration of ^{14}C decreases with the half-life of the decay.

1.1.2   *Units*

The *energy* of radiation is usually measured in units of electronvolt (eV). This unit is defined as the energy gained by an electron when it is accelerated through a potential difference of 1 volt. Multiples of the unit are $1\,\text{keV} = 10^3\,\text{eV}$, $1\,\text{MeV} = 10^6\,\text{eV}$, and $1\,\text{GeV} = 10^9\,\text{eV}$. The relation between the SI unit joule and the electronvolt is:

$$1\,\text{eV} = 1.602 \times 10^{-19}\,\text{J} \tag{1.1}$$

The *rest mass m* of a particle is measured by using the relation $E = mc^2$, in units of eV/c^2. Its relation to the SI unit kg is

$$1\,\text{eV}/c^2 = 1.78 \times 10^{-36}\,\text{kg} \tag{1.2}$$

The *momentum P* of a particle can also be measured in a unit related to the electronvolt. Since the total energy E of the particle is $E^2 = P^2c^2 + m^2c^4$, the unit is eV/c. Again its relation to the SI unit is

$$1\,\text{eV}/c = 0.535 \times 10^{-27}\,\text{kg m/s} \tag{1.3}$$

The *flux I* of particles is defined as

$$I = \frac{n}{tf} \tag{1.4}$$

if n particles penetrate an area f in a time interval t. The unit is particles/(m^2s). The same unit is used for the intensity of a particle beam.

The *activity* of a radioactive source is defined as the number of decays per second; the unit is the curie (Ci), where

$$1\ Ci = 3.7 \times 10^{10}\ decays/s \tag{1.5}$$

or the Becquerel (Bq), where

$$1\ Bq = 1\ decay/s = 2.70 \times 10^{-11}\ Ci \tag{1.6}$$

The activity is related to the *decay constant* λ by the decay law

$$\frac{dN}{dt} = -\lambda N \tag{1.7}$$

such that λ is measured in units of s^{-1}.

The *mean lifetime* τ of a radioactive isotope or particle is defined as the interval after which the initial number N_0 of decaying nuclei or particles has decreased to the value N_0/e. It is related to the decay constant by

$$\tau = \frac{1}{\lambda} \tag{1.8}$$

In nuclear physics, it is customary to use instead of τ the *half-life* $t_{\frac{1}{2}}$, after which half of the initial nuclei have decayed:

$$t_{\frac{1}{2}} = \tau \ln 2 = 0.693\ \tau \tag{1.9}$$

The effects of radiation on matter are measured by three quantities:

(a) the *energy dose D* is defined as the energy W_D absorbed in the material of volume V and density ρ: $D = dW_D/\rho\ dV$. The unit for D is

$$1\ rad = 10^{-2}\ J/kg \tag{1.10}$$

or

$$1\ gray = 1\ J/kg$$

(b) the ion dose D_I is given by the charge Q liberated by the radiation in air of density ρ_A:

$$D_I = \frac{dQ}{\rho_A\ dV} \tag{1.11}$$

The unit of ion dose is the roentgen (R)

$$1\ R = 2.58 \times 10^{-4}\ C/(kg\ of\ air) \tag{1.12}$$

An ion dose of 1 R in air corresponds to a number of $1\ R/e = 1.61 \times 10^{15}$ ions/kg and an energy dose of $1\ R \cdot W_i/e$, where W_i is the mean effective energy needed for liberating one ion–electron pair in air. Since $W_i = 33.7$ eV for air, we obtain the energy dose $D = 0.87$ rad. The number of ion pairs liberated in $1\ cm^3$ of air at standard conditions is 2.08×10^9.

(c) The *equivalent dose* is a measure for the effect of radiation on the human body. It is defined as

$$D_q = qD \tag{1.13}$$

where q is a quality factor for the biological effect of different types of radiation on human tissue. The unit of the equivalent dose is the Rem (roentgen equivalent man), $1\,\text{Rem} = q \cdot 1\,\text{rad}$. More recently, the SI unit sievert (Sv) has been introduced: $1\,\text{Sv} = 100\,\text{Rem}$. The quality factors are approximately $q = 1$ for γ-rays and electrons, $q = 10$ for α-particles, protons and deuterons, $q = 20$ for heavy nuclear fragments and $2 < q < 10$ for neutrons, depending on their energy.

1.1.3 *Artificial radioactivity*

Energetic particles coming from accelerators or nuclear reactors impinging on stable nuclei can undergo nuclear reactions in such a way that unstable radioactive nuclei are produced.

In this way, neutrons from reactors are used to produce β-active isotopes with half-lives between fractions of seconds and 10^5 years. Most β-emitters are also γ-ray sources since the β-decay leads to an excited state of the daughter nucleus, which in turn decays by a γ-transition to the ground state. A few β-decays lead directly to the ground state of the daughter nucleus, and some of these 'pure' β-emitters are listed in table 1.

More useful for calibration purposes are mono-energetic electrons from internal conversion. These are electrons from one of the shells of the atom which are emitted as a consequence of a nuclear de-excitation process if γ-ray emission is suppressed by selection rules. Some examples of such conversion electron sources are ^{137}Cs (625 keV), ^{110m}Ag–^{110}Ag (656 keV, 885 keV) and ^{113m}In (393 keV).

The most common way of decay for an excited nuclear state is the emission of γ-radiation. The excited states are frequently the decay products of a β-transition, such that the half-life of these γ-ray sources is determined

Table 1. *Pure β-sources*

Isotope	End-point energy (keV)	Half-life $t_{\frac{1}{2}}$
^{3}H	18.6	12.26 years
^{14}C	156	5730 years
^{33}P	248	24.4 days
^{90}Sr	546	27.7 years
^{90}Y	2.27×10^3	64 hr
^{99}Tc	292	2.1×10^5 years

by the half-life of the β-decay. A few examples of such nuclides are listed in table 2.

Mono-energetic photon radiation in the keV range can also be obtained from X-ray transitions in the atomic shell. A very useful example for such a source is ^{55}Fe with an energy of 5.9 keV from the K_α X-ray transition in manganese. This energy corresponds to the ionization energy loss of a minimum ionizing charged particle along a path of a few centimetres in gases at standard conditions. The photoelectrons liberated by a photon of this energy can therefore be used for calibration purposes.

Another decay mode useful for calibration purposes is α-decay. The reaction is $^A_Z X \rightarrow ^{A-4}_{Z-2}X + \alpha$, and the decay rate is determined by the tunnelling probability of the α-particle through the potential barrier of the remaining nucleus. This mechanism implies an exponential dependence of the decay rate from the energy E_α of the α-particle emitted. Half-lives range from 10^{10} y at $E_\alpha = 4$ MeV to days at $E_\alpha = 6.5$ MeV. For laboratory purposes, an intermediate half-life is indicated. A useful source is, e.g., ^{241}Am with a half-life of 433 days and two α-lines at 5.49 MeV and 5.44 MeV.

1.1.4 *Particle accelerators*

The study of atomic nuclei and of their constituents and the search for small or point-like building blocks of these constituents required the study of scattering and annihilation processes at ever-larger centre-of-mass energies. This was achieved by constructing particle accelerators. For protons the electrostatic van-de-Graaf accelerator, the weakly focusing cyclotron and the synchro-cyclotron gave accelerated particles up to 15 MeV, 20 MeV and 500 MeV energy in the laboratory system, respectively. In 1956, E. D. Courant and H. A. Snyder invented the principle of strong focusing by alternating magnetic field gradients along the circular orbit. This breakthrough enabled the construction of the proton synchro-

Table 2. γ-ray sources

Parent nucleus of β-decay	$t_{\frac{1}{2}}$		Daughter nucleus	E_γ (keV)
^{22}Na	2.6	years	^{22}Ne	1274
^{57}Co	272	days	^{57}Fe	14.4
				122.1
^{60}Co	5.27	years	^{60}Ni	1173.2
				1332.5
^{137}Cs	30.0	years	^{137}Ba	661.6

tron at the European Laboratory for Particle Physics CERN at Geneva, and of the alternating gradient synchrotron (AGS) at the Brookhaven National Laboratory near Upton (New York). Both accelerators achieved proton energies around 30 GeV. An extension of this principle led to the construction of the proton synchrotron at the Serpukhov Laboratory in the USSR, with a peak energy of 70 GeV. Energies of 400 GeV were reached with two large proton synchrotrons in the 1970s, one at the Fermi National Laboratory (Fermilab) near Batavia (Illinois) and the other at CERN: the super proton synchrotron (SPS). At Fermilab, the development of super-conducting pulsed dipole magnets with a peak magnetic field of 4.5 T was the basis for the tevatron machine in the ring of the existing proton synchrotron. The peak energy of protons of mass m_p in the tevatron is now $E = 800$ GeV, corresponding to a centre-of-mass energy of $\sqrt{s} = \sqrt{(2m_p E)} = 40$ GeV.

At CERN, another line of development was followed in order to increase the centre-of-mass energy in hadron collisions: the SPS was used as a storage ring for protons and antiprotons. However, the collision rate in this storage ring would not have been sufficiently high for the study of such rare processes as the production of weak intermediate bosons if it was not possible to increase the flux of stored antiprotons. The invention and implementation of stochastic cooling of antiproton beams by S. van der Meer solved this problem. At the time of writing, this $\bar{p}p$-collider reaches the highest centre-of-mass energy of all accelerators, $\sqrt{s} = 630$ GeV. A similar proton–antiproton collider will be operational at Fermilab in 1986.

The development of electron accelerators started with the betatron, reaching 45 MeV, and ended with strongly focusing electron synchrotrons at Bonn, Harvard and Cornell Universities and at the German Synchrotron Laboratory DESY (Hamburg) and the two-mile linear accelerator (SLAC), where the peak energy is 30 GeV.

Also in electron accelerators the principle of colliding intersecting beams has been used to increase the centre-of-mass energy of collisions. Initial research on electron–positron storage rings was done at Novosibirsk, Frascati and Cambridge, and the first great successes with this technique came with the SPEAR-ring at Stanford at centre-of-mass energies up to 4 GeV. The second ring in this energy domain was DORIS at DESY, Hamburg. A tenfold increase in centre-of-mass energy was achieved with the PETRA ring at DESY and the similar PEP machine at Stanford. At present, the highest centre-of-mass energies of all electron–positron colliders are reached in the PETRA machine, where $\sqrt{s} = 45$ GeV. A larger machine, called TRISTAN, is being constructed at the Japanese high-energy laboratory KEK, Tsukuba, designed for $\sqrt{s}$ up to 60 GeV.

Construction is also under way at Stanford for a collider fed by the linear accelerator (SLC) and at CERN for the large electron–positron collider (LEP). This circular machine will reach $\sqrt{s} = 100$ GeV at the end of 1988. A list of accelerators in operation at this time is given in table 3.

Table 3. *Large particle accelerators in operation*

Machine	Laboratory	Maximum beam energy (GeV)	Start of operations
Proton synchrotrons			
PS	CERN, Geneva, Switzerland	28	1960
AGS	Brookhaven, Upton, USA	33	1960
PS	Serpukhov, Protvino, USSR	76	1967
SPS	CERN, Geneva, Switzerland	450	1976
Tevatron	Fermilab, Chicago, USA	1000	1985
PS	KEK, Tsukuba, Japan	8	1976
Electron accelerators			
Electron synchrotron	DESY, Hamburg, Germany	7.4	1964
Lin. acc.	SLAC, Stanford, USA	32	1966
Proton–antiproton storage rings			
SPPS	CERN, Geneva, Switzerland	450	1982
Collider	Fermilab, Chicago, USA	1000	1986
Electron–positron storage rings			
ADONE	Frascati, Italy	1.5	1969
DCI	LAL, Orsay, France	1.8	1976
SPEAR	SLAC, Stanford, USA	4	1972
DORIS	DESY, Hamburg, Germany	5	1974
VEPP IV	Novosibirsk, USSR	7	1980
CESR	Cornell, USA	8	1979
PETRA	DESY, Hamburg, Germany	22	1978
PEP	SLAC, Stanford, USA	18	1980
SLC	SLAC, Stanford, USA	50	1986
LEP	CERN, Geneva, Switzerland	50	1988
Heavy ion accelerators			
UNILAC	GSI, Darmstadt, Germany	20 MeV/nucleon (U)	1976
GANIL	Caen, France	95 MeV/nucleon (O)	1983
		60 MeV/nucleon (Ca)	1983
		35 MeV/nucleon (Kr)	1983
Super-HILAC	LBL, Berkeley, USA	9.5 MeV/nucleon (all nuclei)	1970
Medium energy proton accelerators – pion factories			
LAMPF	LANL, Los Alamos, USA	800 MeV	1973
SIN	SIN, Villigen, Switzerland	590 MeV	1974
TRIUMF	TRIUMF, Vancouver, Canada	520 MeV	1974

1.2   Interactions of particles and γ-radiation with matter

The physical processes which enable us to detect particles are different for neutral and charged particles. Photons can interact by the photoelectric or Compton effects or by creation of an electron–positron pair; the latter process dominates at energies above 5 MeV. The electrons or positrons resulting from these interactions can be detected in the same way as other charged particles. Neutrons interact strongly with nuclei, and in doing so they produce charged secondary particles. Neutrinos can only be detected by their weak interaction with nuclei or with electrons: lepton number conservation requires the emission of a charged or neutral lepton in these processes, and in addition hadrons are created in inelastic reactions.

Charged particles can be detected directly through their electromagnetic interactions with the atomic electrons of the detector material. This is treated in section 1.2.1, and the interactions of γ-rays are described in section 1.2.2.

1.2.1   *Detection of charged particles*

For the detection of charged particles, use is made of their electromagnetic interaction. If a charged particle traverses a layer of material, three processes can occur: atoms can be ionized, the particle can emit Cherenkov radiation, or the particle can cause the emission of transition radiation. A unified deduction of the energy loss by ionization and of the intensity of the radiation emitted can be found in the work of Allison and Wright [AL 83b, AL 80]. Consider the electromagnetic interaction of a charged particle of mass M and velocity $v = \beta c$ in a material of refractive index n and with the dielectric constant $\varepsilon = \varepsilon_1 + i\varepsilon_2$, such that $\varepsilon_1 = n^2$. In the interaction, a photon of energy $\hbar\omega$ and momentum $\hbar k$ is created. Energy momentum conservation gives a relation between the four-momenta of the incoming particle (P), the outgoing particle (P') and the photon (P_γ): $P' = P - P_\gamma$. For small photon energies $(\hbar\omega \ll \gamma Mc^2)$ this gives

$$\omega = \mathbf{v} \cdot \mathbf{k} = vk \cos \theta_c \qquad \vec{P} = \vec{P}_\gamma + \vec{P}' \;\Rightarrow\; \vec{P}' = \vec{P} - \vec{P}_\gamma \qquad (1.14)$$

where θ_c is the angle between the directions of the emitted photon and the incoming particle. In a material the photon energy and momentum are related by the dispersion relation

$$\omega^2 = \frac{k^2 c^2}{\varepsilon} \qquad (1.15)$$

Eqs. (1.14) and (1.15) yield

$$\sqrt{\varepsilon}\,\frac{v}{c} \cos \theta_c = 1 \qquad (1.16)$$

At photon energies below the excitation energies of the material ('optical region') ε is real and $\varepsilon > 1$ such that θ_c is real for $v > c/\sqrt{\varepsilon}$. The emission of real photons is then possible ('Cherenkov effect') if the velocity of the particle is larger than the phase velocity $c/\sqrt{\varepsilon}$ of light in the material ('Cherenkov threshold'). At photon energies in the range from 2 eV through 5 keV, $\varepsilon = \varepsilon_1 + i\varepsilon_2$ is a complex number with $\varepsilon_2 > 0$ and $\varepsilon_1 < 1$. In this case, only virtual photons are exchanged between the particle and the atoms of the material, resulting in excitation or ionization of the atoms and a corresponding energy loss of the particle. Finally, in the X-ray domain, i.e. at photon energies above 5 keV, the absorption coefficient becomes small ($\varepsilon_2 \ll 1$), and still $\varepsilon_1 < 1$. The threshold velocity for the Cherenkov effect is then larger than the light velocity *in vacuo*. In spite of this, radiation is emitted below this threshold if there are discontinuities in the material traversed by the particle. This is called transition radiation.

In a simplified two-dimensional model [AL 83b], some properties of these processes can be considered: let the particle move along the z-axis of the co-ordinate system (i.e. $\mathbf{v} = (0, 0, v)$), and the observer be situated at the point $(0, y, z)$. Then from eq. (1.14) $vk_z = \omega$ and, from eq. (1.15), $k_y^2 + k_z^2 = \omega^2 \varepsilon / c^2$. This means that

$$k_y = \frac{\omega}{v} \sqrt{\left(\frac{v^2 \varepsilon}{c^2} - 1 \right)}$$

If we call the phase velocity of the light in the material $c_m = c/\sqrt{\varepsilon}$, $\beta' = v/c_m$ and $\gamma' = 1/\sqrt{(1 - \beta'^2)}$, then

$$k_y = \frac{\omega}{v} \sqrt{(\beta'^2 - 1)} \tag{1.17}$$

We can distinguish two cases:

(a) $\beta' > 1$: k_y and k_z are real, a real wave of the form $\exp(i(\mathbf{kr} - \omega t))$ is created.

(b) $\beta' < 1$: k_y is purely imaginary, the transverse y-component of the electromagnetic field does not oscillate but is damped aperiodically with the attenuation length y_0:

$$\exp[i(\mathbf{kr} - \omega t)] = \exp\left[i \frac{\omega}{v} (z - vt) \right] \cdot \exp(-y/y_0) \tag{1.18}$$

with

$$y_0 = \frac{v}{\omega} \frac{1}{\sqrt{(1 - \beta'^2)}} = \frac{\beta' \gamma'}{k} \tag{1.19}$$

The range of the transverse field increases linearly with $\beta' \gamma'$ if the velocity of the particle increases. This relativistic expansion of the transverse field causes the relativistic rise of the energy loss by ionization. Expressing the range y_0 from eq. (1.19) in the variables β and γ gives

$$y_0 = \frac{\beta}{\sqrt{[1/\gamma^2 + (1-\varepsilon)\beta^2]k_0}} \qquad (1.20)$$

with k_0 being the wave vector *in vacuo*. There are two possibilities:

(i) $\varepsilon > 1$. This is true for photon energies in the optical region below the excitation of the medium; if β' approaches unity from below, the range of the transverse field increases until the Cherenkov threshold $\beta' = 1$, where $y_0 \to \infty$.

(ii) $\varepsilon < 1$. This is valid above the ionization threshold. The transverse range increases but the denominator in eq. (1.20) does not vanish, such that y_0 reaches a maximum value for $\beta \to 1$ and $\gamma \to \infty$

$$y_0^{max} = \frac{1}{k\sqrt{(1-\varepsilon)}} \qquad (1.21)$$

This plateau in y_0 corresponds to a saturation in the energy loss by ionization; this is reached at a velocity at which the two terms in the denominator of eq. (1.20) are equal, i.e. at

$$(\beta\gamma)_{sat} \sim \frac{1}{\sqrt{(1-\varepsilon)}} \qquad (1.22)$$

Since the dielectric susceptibility $\varepsilon - 1$ is proportional to the density ρ of the medium, saturation is reached earlier for dense materials:

$$(\beta\gamma)_{sat} \sim \frac{1}{\sqrt{\rho}} \qquad (1.23)$$

This density effect can therefore be understood as a consequence of the transverse field reaching neighbour atoms: for a high-density medium, this occurs at lower particle velocities than in gases. A large relativistic increase of 50–70% is observed in noble gases at standard conditions, while for solids and liquids the increase amounts only to a few per cent.

The exact calculation in the photo absorption model [AL 80] gives the differential cross-section per electron and per energy loss dE of the charged particle:

$$\frac{d\sigma}{dE} = \frac{\alpha}{\beta^2\pi} \frac{\sigma_\gamma(E)}{EZ} \ln\frac{1}{\sqrt{[(1-\beta^2\varepsilon_1)^2 + \beta^4\varepsilon_2^2]}}$$
$$+ \frac{\alpha}{\beta^2\pi} \frac{\sigma_\gamma(E)}{EZ} \ln\left(\frac{2mc^2\beta^2}{E}\right)$$
$$+ \frac{\alpha}{\beta^2\pi} \frac{1}{E^2} \int_0^E \frac{\sigma_\gamma(E')}{Z} dE' + \frac{\alpha}{\beta^2\pi} \frac{1}{ZN\hbar c}\left(\beta^2 - \frac{\varepsilon_1}{|\varepsilon|^2}\right)\theta \qquad (1.24)$$

Here $\alpha = e^2/(4\pi\varepsilon_0\hbar c) = 1/137$ is the fine structure constant; $\varepsilon = \varepsilon_1 + i\varepsilon_2$ is the complex dielectric constant; θ is the phase of the complex expression $1 - \varepsilon_1\beta^2 + i\varepsilon_2\beta^2$; σ_γ is the cross-section for absorption of a photon of energy E by the atoms of the medium; and $N = N_0\rho/A$ is the atomic density.

For photon energies below the excitation energy of the atom, i.e. in the optical region where σ_γ vanishes, the only remaining term is the fourth one. It describes the production of Cherenkov radiation. In this region $\varepsilon_2 = 0$ and $\varepsilon = \varepsilon_1$, such that the phase θ of the expression $1 - \varepsilon_1\beta^2$ vanishes below the Cherenkov threshold ($\beta^2 = 1/\varepsilon_1$) and jumps to π above the threshold. Multiplying the term with the density of electrons yields above the threshold the flux of photons in the energy interval $dE = \hbar\, d\omega$ from a path length L of the charged particle in the medium:

$$\frac{dn}{\hbar\, d\omega} = \frac{\alpha}{\hbar c}\left(1 - \frac{1}{\beta^2\varepsilon}\right)L \tag{1.25}$$

For long radiators ($L \gg \lambda$) this gives the intensity of Cherenkov radiation; for short radiators the relation can be used to derive, for small emission angles ϕ and β near unity, the number of transition radiation photons in the X-ray domain. Here ε is close to unity, $\varepsilon = 1 - \omega_p^0/\omega^2$, where $\omega_p = \sqrt{(ZNe^2/\varepsilon_0 m_e)}$ is the 'plasma frequency' of electrons. The number of such photons per interval of frequency, $d\omega$, and of solid angle, $d\Omega = 2\pi d(\cos\phi)$, is:

$$\frac{d^2n}{d\omega\, d\Omega} = \frac{\alpha}{\pi^2\omega}\,\phi^2\, 4\sin^2\left[\frac{\omega L}{4c}\left(\frac{\omega_p^2}{\omega^2} + \phi^2 + \frac{1}{\gamma^2}\right)\right]$$
$$\times\left(\frac{1}{1/\gamma^2 + \omega_p^2/\omega^2 + \phi^2} - \frac{1}{1/\gamma^2 + \phi^2}\right)^2 \tag{1.26}$$

This transition radiation arises from the interference of Cherenkov emission at the two boundary surfaces of the foil of thickness L with different phase. The intensity of photons emitted has its maximum around the polar angle $\phi \sim 1/\gamma$. Eq. (1.26) can be interpreted as the intensity from two amplitudes A and $-A\exp(i\delta)$ originating from the entry and exit faces of the thin foil, yielding $|A|^2\, 4\sin^2(\delta/2)$. With random phases δ in the interval $0 \leqslant \delta \leqslant 2\pi$ this would yield $2|A|^2$. Under this (invalid) assumption one can integrate the intensity from *one* discontinuity over the solid angle $d\Omega = 2\pi\phi\, d\phi$ and obtain, for $\omega \ll \omega_p$:

$$\frac{dn}{d\omega} \sim \frac{2\alpha}{\pi\omega}\ln\left(\frac{\gamma\omega_p}{\omega}\right) \tag{1.27}$$

and an energy flux proportional to $\gamma\alpha\hbar\omega_p$. In reality, the modulation of the radiation intensity by the interference term $4\sin^2(\delta/2)$ cannot be avoided. The term is maximum for $\delta = n\pi$ ($n = 1, 3, 5, \ldots$). If we define a formation zone Z:

$$Z = \frac{2c}{\omega} \left(\frac{1}{\gamma^2} + \frac{\omega_p^2}{\omega^2} + \phi^2 \right)^{-1}$$

then the modulation term is just $4\sin^2(L/2Z)$ and the first saturation value for the intensity is reached at $L/Z \sim \pi$. For the average values $\phi \sim 1/\gamma$, $\omega \sim \gamma \omega_p$, Z is approximately $\lambda \gamma^2/(3\pi)$, where $\lambda = 2\pi c/\omega$ is the wavelength of the radiation. The formation zone has to be within the foil (so $Z < L$), i.e. the radiation intensity does not increase linearly with γ but approaches a saturation value as γ approaches the value $\gamma_s = \sqrt{(3L/\lambda)}$. Counters detecting Cherenkov or transition radiation are discussed in chapter 5.

The first three terms in eq. (1.24) are responsible for the energy loss of a charged particle by ionization. The third one describes the probability for generating an energetic knock-on electron ('δ-ray'). From the other two terms the differential average energy loss dE/dx per path length dx can be obtained. This is done by integrating the transferred energy E from a value corresponding to the average ionization potential I of the atom up to the maximum energy of a struck electron $(2m_e c^2 \beta^2 \gamma^2)$. This approximation is the Bethe–Bloch formula [BE 30, BE 32, BE 33, ST 71]:

$$-\frac{dE}{dx} = (4\pi r_e^2 m_e c^2 N_0 Z z^2/A\beta^2) \left\{ \ln\left[\frac{2m_e c^2 \beta^2}{(1-\beta^2)I} \right] - \beta^2 \right\} \qquad (1.28)$$

where N_0 is Avogadro's number; Z and A are the atomic number and mass number of the material transversed; ze and $v = \beta c$ are the charge and velocity of the ionizing particle; m_e is the electron mass; $r_e = 2.8$ fm, the classical electron radius; and I is the effective ionization potential. Instead of the thickness x of the material with density ρ, often the mass per area $X = \rho x$ is used and measured in units of g/cm^2. Then

$$\frac{dE}{dX} = \frac{1}{\rho} \frac{dE}{dx} \qquad (1.29)$$

In eq. (1.28) the density effect discussed above (eq. (1.23)) is neglected, i.e. the relation is valid for low-density materials, such as gases at standard conditions. The values of the effective ionization potential can be approximated by $I = I_0 Z$, with $I_0 = 12$ eV, or by the Thomas–Fermi model which gives $I = I_1(Z)^{0.9}$, with $I_1 = 16$ eV. Precise values for I_0 can be found in table 4. They are close to the minimal ionization energies for the atoms in question. Table 4 also contains values for the minimal excitation energy of the atom or molecule, E_{ex}.

The energy loss dE/dx does not depend on the mass m of the ionizing particles, but on their velocity $v = \beta c$. As a function of β, dE/dx at low β decreases as $1/\beta^2$, then reaches a minimum around $\beta\gamma = P/mc \sim 4$, and increases for relativistic velocities with $\beta\gamma \to \infty$. The ratio of the asymptotic

Table 4. *Properties of gases at normal conditions: density ρ, minimal energy for ionization E_i, minimal energy for excitation E_{ex}, minimal energy for ionization potential per atomic electron $I_0 = I/Z$, energy loss W_i per ion pair produced, minimal energy loss $(dE/dx)_0$, total number of ion pairs n_T and number of primary ions n_p per cm path for minimum ionizing particles* [SA 77]

Gas	Z	A	ρ (g/cm³)	E_{ex} (eV)	E_i (eV)	I_0 (eV)	W_i (eV)	$(dE/dx)_0$ (MeV g cm⁻²)	$(dE/dx)_0$ (keV/cm)	n_P (cm)⁻¹	n_T (cm)⁻¹
H_2	2	2	8.38×10^{-5}	10.8	15.9	15.4	37	4.03	0.34	5.2	9.2
He	2	4	1.66×10^{-4}	19.8	24.5	24.6	41	1.94	0.32	5.9	7.8
N_2	14	28	1.17×10^{-3}	8.1	16.7	15.5	35	1.68	1.96	(10)	56
O_2	16	32	1.33×10^{-3}	7.9	12.8	12.2	31	1.69	2.26	22	73
Ne	10	20.2	8.39×10^{-4}	16.6	21.5	21.6	36	1.68	1.41	12	39
Ar	18	39.9	1.66×10^{-3}	11.6	15.7	15.8	26	1.47	2.44	29.4	94
Kr	36	83.8	3.49×10^{-3}	10.0	13.9	14.0	24	1.32	4.60	(22)	192
Xe	54	131.3	5.49×10^{-3}	8.4	12.1	12.1	22	1.23	6.76	44	307
CO_2	22	44	1.86×10^{-3}	5.2	13.7	13.7	33	1.62	3.01	(34)	91
CH_4	10	16	6.70×10^{-4}		15.2	13.1	28	2.21	1.48	16	53
C_4H_{10}	34	58	2.42×10^{-3}		10.6	10.8	23	1.86	4.50	(46)	195

value of dE/dx at highest momenta to the minimal value amounts to 1.5 for gases at standard conditions. In denser materials, such as gases at high pressures, liquids and solids, the relativistic rise is much smaller. The reason for this density effect was discussed in eq. (1.23). Fig. 1.1 shows the dependence of dE/dx from $\beta\gamma$ for a mixture of argon and 5% methane, together with experimental data on the relativistic rise [LE 78a]. The minimal value of the energy loss, $(dE/dx)_0$, at $\beta\gamma \sim 4$ amounts to approximately 1–2 MeV/g cm^2 (see table 4).

Eq. (1.28) gives the average energy loss of a particle. However, the differential distribution of energy losses is not, in general, a gaussian distribution. In order to obtain this distribution, the following procedure is used [LA 44, AL 80, BI 75]: let $\sigma = \int (d\sigma/dE)\, dE$ be the total cross-section for the collisions between the moving particle and the atomic electrons. Then σNx is the probability for interaction in the path length x. If we subdivide this length into n intervals such that the probability for *one* collision becomes $\alpha = \sigma N(x/n) \ll 1$, then the energy loss distribution over this small interval x/n is

$$f\left(\frac{x}{n}, E\right) = \left(1 - \sigma N \frac{x}{n}\right)\delta(E) + N \frac{x}{n}\frac{d\sigma}{dE}(E) \qquad (1.30)$$

Fig. 1.1. Energy loss by ionization, normalized to the minimal value $(dE/dx)_0$ at $\beta\gamma = 4$, for an argon–methane (5%) mixture. Measured points after [LE 78a]; dashed curve calculation after [ST 52]; dashed–dotted curve [ER 77]; full line, photo absorption model for ionization [CO 75, CO 76, AL 80].

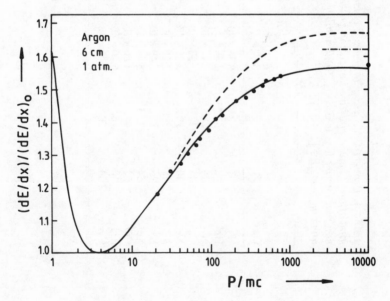

The first term is the probability for no collision multiplied with an energy loss distribution peaking at $E = 0$; the second one is the probability for one collision multiplied with the cross-section for collisions with the energy loss E. Adding two such thin layers, the energy loss distribution becomes

$$f\left(\frac{2x}{n}, E\right) = \int_0^E f\left(\frac{x}{n}, E\right) \cdot f\left(\frac{x}{n}, E - \varepsilon\right) d\varepsilon \tag{1.31}$$

By iterating this procedure n times, the energy loss in the layer of thickness x is obtained.

The result of the calculation then depends entirely on the form of $d\sigma/dE$ used for the model. An oversimplified approach would be the assumption that only one energy transfer E^* is possible. Then

$$\frac{d\sigma}{dE} = \sigma \delta(E - E^*)$$

For this form of the cross-section

$$f\left(\frac{x}{n}, E\right) = (1 - \alpha)\delta(E) + \alpha\delta(E - E^*)$$

$$f\left(\frac{2x}{n}, E\right) = (1 - \alpha)^2\delta(E) + 2(1 - \alpha)\alpha\delta(E - E^*) + \alpha^2\delta(E - 2E^*)$$

For n collisions in the n layers of thickness x/n the energy loss distribution is

$$f(x, E) = \sum_{\nu=0}^n \binom{n}{\nu} \alpha^\nu (1 - \alpha)^{n - \nu} \delta(E - \nu E^*)$$

For n very large and $\alpha < 10^{-3}$, only the first few terms contribute, such that

$$f(x, E) = \sum_{\nu = 0, 1, 2} \frac{\alpha^\nu}{\nu!} e^{-\alpha} \delta(E - E^*) \tag{1.32}$$

which is the sum of Poisson distributions.

This example already demonstrates one property of energy loss distributions, remarked on first by Landau [LA 44]: they are asymmetric with a tail at high values which is caused by collisions at small impact parameters and, consequently, large energy transfers ('δ-rays').

For the actual calculation, different models have been used: Landau's [LA 44] and Sternheimer's [ST 52] approaches are based on Rutherford scattering of the moving particle of electrons with the atomic binding energy approximated by an average ionization potential; Blunck and Leisegang [BL 50] include approximate atomic shell corrections; and Allison and Cobb [AL 80] refine this procedure by including all details of the atomic shell structure.

Fig. 1.2 shows that the earlier calculations do not compare well with the measured energy loss distribution in a 1.5 cm-thick layer of argon at STP

[HA 73]. However, the newest models ('photo absorption model', PAI [CO 75, CO 76, AL 80]) give a satisfactory description of the data. These calculations also reproduce the size of the measured relativistic rise as shown in fig. 1.1, while in earlier models [ST 52, ER 77] the rise was too large by 10–15%.

The following picture emerges about the process of energy loss by ionization and excitation: in the primary process atoms are excited and ionized – the energy distribution of the ejected electrons is proportional to $1/E^2$. Electrons with energies above 100 eV are able to ionize further atoms in secondary collisions. The total number of liberated ions n_T is proportional to the energy loss ΔE of the fast moving particle

$$n_T = \frac{\Delta E}{W_i} \tag{1.33}$$

where W_i is the energy loss per ion pair produced. This total number n_T is two to seven times larger than the number of primary ion pairs, n_p. Table 3 contains values of W_i, n_T and n_p for several gases. The average energy W_i needed for creating one electron–ion pair in gases lies between 41 eV in helium and 22 eV in xenon.

In semiconductors this energy amounts only to 3.5 eV in silicon and

Fig. 1.2. Differential distribution of energy losses in a layer of 1.5 cm argon with 5% methane at STP. Histogram: data for π mesons and electrons of 3 GeV/c momentum [HA 73]. Model calculations: dashed curve [LA 44, MA 69]; dotted curve [LA 44, BL 50]; full curve [AL 80].

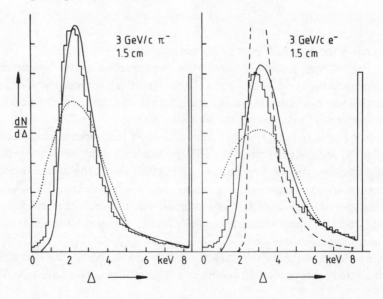

2.85 eV in germanium; hence, the number of ions created is much larger for the same energy deposition. Consequently, the statistical fluctuations in this number are much smaller, and counters based on semiconductors have a very good relative energy resolution ($\Delta E/E \sim 10^{-3}$–10^{-4}). On the other hand, the fabrication of large Si or Ge crystals of the purity required for this application meets technical difficulties, and therefore the use of such semiconductor counters is restricted to nuclear physics applications.

In liquid noble gases the values of W_i are near to those for the gaseous phase: $W_i(\text{L Ar}) = 23.6$ eV and $W_i(\text{L Xe}) = 16$ eV. Their high density compared with gases makes the use of such liquids for ionization detectors in total absorbing shower counters and calorimeters attractive. The collection of charges in such materials requires, however, a very high degree of purity of the liquid.

1.2.2 Detection of γ-rays

If a beam of photons with intensity I_0 traverses a layer of material of thickness x or of a mass thickness $X = \rho x$, the intensity emerging from the layer is

$$I(X) = I_0 \, e^{-\mu x} = I_0 \, e^{-(\mu/\rho)X} \tag{1.34}$$

Here μ is called the linear absorption coefficient and μ/ρ the mass absorption coefficient. It is related to the cross-section σ for photon absorption by $\mu = \sigma N_0 \rho / A$, where N_0 is Avogadro's constant; ρ is the density of the material; and A is the mass of a mole of the material. The main contributions to this cross-section at photon energies above 10 keV are: the photoelectric effect ($\gamma + \text{atom} \rightarrow \text{ion} + e^-$) in which all the energy of the photon is transferred to a bound atomic electron, and which dominates at energies below 100 keV; the Compton effect ($\gamma + e^- \rightarrow \gamma + e^-$), where the photon scatters off a quasi-free electron whose binding energy can be neglected (this process is important at energies around 1 MeV); and pair creation ($\gamma + \text{nucleus} \rightarrow e^+ + e^- + \text{nucleus}$) in which an electron–positron pair is created in the vicinity of a nucleus and which prevails for $E_\gamma > 2$ MeV. Fig. 1.3 shows the dependence of the mass absorption coefficient μ for lead from the photon energy E_γ. It appears from this figure that the contribution from the photoelectric effect increases sharply at the energy corresponding to the binding energy of electrons in a particular atomic shell; the edges due to the L- and K-shells can be seen in this figure. Between the binding energies of the L- and K-shells and above the K-edge, the absorption coefficient falls approximately as $1/E_\gamma^3$, while the contribution from the Compton effect decreases with $1/E_\gamma$. The part of the absorption coefficient due to pair creation rises steeply above the threshold at $E_\gamma = 2m_e = 1.02$ MeV.

The cross-section corresponding to these processes can be calculated. The results can be approximated by the following relations as a function of the reduced photon energy $\varepsilon = E_\gamma / m_e c^2$:

Photoelectric absorption
In the energy range between the K-absorption edge ε_K and $\varepsilon = 1$ the following approximation holds:

$$\sigma_{\text{ph}} = \frac{32\pi}{3} \sqrt{2} \, Z^5 \alpha^4 \frac{1}{\varepsilon^{\frac{7}{2}}} r_e^2 \quad \text{for } \varepsilon_K < \varepsilon < 1 \tag{1.35}$$

and for $\varepsilon > 1$:

$$\sigma_{\text{ph}} = 4\pi r_e^2 Z^5 \alpha^4 \frac{1}{\varepsilon} \tag{1.36}$$

Compton effect
Initially the electron is at rest; its four-momentum is $P_e = (m_e c^2, \mathbf{0})$ and the

Fig. 1.3. Mass absorption coefficient μ/ρ for photons in lead.

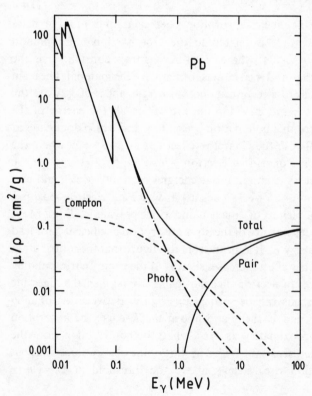

photon four-momentum $P_\gamma = (0, \mathbf{P}_\gamma)$. If we call the quantities after the collision P'_e and P'_γ, four-momentum conservation gives

$$P_e + P_\gamma = P'_e + P'_\gamma \tag{1.37}$$

and from this follows the energy $E'_\gamma = |P'_\gamma|$ of the scattered photon

$$E'_\gamma = \frac{E_\gamma}{1 + (E_\gamma/m_e c^2)(1 - \cos\theta)} \tag{1.38}$$

where θ is the scattering angle of the photon. The energy difference between the photon after and before scattering is transferred to the electron, which then has the kinetic energy

$$T'_e = \frac{E_\gamma^2}{m_e c^2} \frac{1 - \cos\theta}{1 + (E_\gamma/m_e c^2)(1 - \cos\theta)} \tag{1.39}$$

Two extreme cases can be considered:

(a) Scattering at very small angles, $\theta \sim 0$: then $E'_\gamma \sim E_\gamma$ and $T'_e \sim 0$.

(b) Backscattering at $\theta = \pi$: then

$$E'_\gamma = \frac{E_\gamma}{1 + 2(E_\gamma/m_e c^2)} \to \frac{m_e c^2}{2} \quad \text{for } E_\gamma \gg m_e c^2$$

and the maximum electron energy ('Compton edge') is

$$T'_e = E_\gamma \frac{2\varepsilon}{1 + 2\varepsilon} \to E_\gamma \left(1 - \frac{1}{2\varepsilon}\right) \quad \text{for } E_\gamma \gg m_e c^2$$

The electron recoil spectrum visible in the detector is a continuum between zero and the Compton edge. The gap between the Compton edge and the energy of the incident γ-ray becomes a constant, $m_e c^2/2 = 0.256$ MeV; in the limiting case $E_\gamma \gg m_e c^2$.

The total scattering cross-section per electron at very low energies is given by the classical Thomson formula

$$\sigma_{Th} = \frac{8\pi}{3} r_e^2 = 0.665 \text{ barn}$$

For relativistic photon energies, the quantum-mechanical calculation leads to the Klein–Nishina cross-section formula:

$$\sigma_c = 2\pi r_e^2 \left\{ \left(\frac{1+\varepsilon}{\varepsilon^2}\right) \left[\frac{2(1+\varepsilon)}{1+2\varepsilon} - \frac{1}{\varepsilon} \ln(1+2\varepsilon)\right] + \frac{1}{2\varepsilon} \ln(1+2\varepsilon) - \frac{1+3\varepsilon}{(1+2\varepsilon)^2} \right\} \tag{1.40}$$

For two extreme cases the relation becomes more transparent; for $\varepsilon \ll 1$ we obtain

$$\sigma_c = \sigma_{Th}(1 - 2\varepsilon)$$

and for $\varepsilon \gg 1$

$$\sigma_c = \frac{3}{8} \sigma_{Th} \frac{1}{\varepsilon} \left(\frac{1}{2} + \ln 2\varepsilon\right)$$

The angular distribution of scattered γ-rays is forward–backward symmetric for Thomson scattering

$$\frac{d\sigma}{d\Omega} = \frac{1}{2} r_e^2 (1 + \cos^2 \theta) \tag{1.41}$$

For relativistic Compton scattering, the angular distribution becomes asymmetric with a forward peak as given by the differential Klein–Nishina formula:

$$\frac{d\sigma}{d\Omega} = \frac{1}{2} r_e^2 \left(\frac{1}{1 + \varepsilon(1 - \cos\theta)} \right)^3 [-\varepsilon \cos^3 \theta + (\varepsilon^2 + \varepsilon + 1)(1 + \cos^2 \theta)$$
$$- \varepsilon(2\varepsilon + 1) \cos\theta] \tag{1.42}$$

Pair production cross-section per nucleus
For $1 < \varepsilon < 137/Z^{\frac{1}{3}}$

$$\sigma_p = r_e^2 4\alpha Z^2 (\tfrac{7}{9} \ln 2\varepsilon - \tfrac{109}{54}) \tag{1.43}$$

For $\varepsilon \gg 137/Z^{\frac{1}{3}}$

$$\sigma_p = r_e^2 4\alpha Z^2 \left[\frac{7}{9} \ln \left(\frac{183}{Z^{\frac{1}{3}}} \right) - \frac{1}{54} \right] \tag{1.44}$$

The mass absorption coefficient for pair creation, $\mu_p/\rho = \sigma_p N_0/A$, reaches, for high photon energies, an asymptotic value μ_p^0/ρ which is obtained from eq. (1.44) by neglecting the last term:

$$\mu_p^0 = r_e^2 4\alpha Z^2 \frac{N_0}{A} \frac{7}{9} \ln \frac{183}{Z^{\frac{1}{3}}} = : \frac{7}{9} \frac{1}{X_0} \tag{1.45}$$

The functional dependence of μ_p on E_γ is shown in fig. 1.4 for some materials. The 'radiation length' X_0 defined in eq. (1.45) corresponds to a layer thickness of material where pair creation happens with a probability $P = 1 - e^{-\frac{7}{9}} \approx 54\%$ at high photon energies. Table 5 contains values of X_0 for a few materials.

1.2.3   *Bremsstrahlung of electrons*

At energies far above 1 MeV, the energy loss by ionization for fast electrons has the form ($\beta \sim 1$)

$$-\left(\frac{dE}{dx} \right)_{\text{ion}} = 4\pi N_0 \frac{Z}{A} r_e^2 m_e c^2 [\ln(2mv^2\gamma^2/I) - 1] \tag{1.46}$$

Electrons of high energy, because of their low mass, can also lose energy by a second process, namely by radiating photons while being decelerated in the Coulomb field of a nucleus ('Bremsstrahlung'). The average energy loss by Bremsstrahlung in traversing a layer of thickness dx is calculated to be

$$-\left(\frac{dE}{dx} \right)_{\text{Brem}} = 4\alpha N_0 \frac{Z^2}{A} r_e^2 E \ln \frac{183}{Z^{\frac{1}{3}}} = \frac{E}{X_0} \tag{1.47}$$

where again the radiation length X_0 defined in section 1.2.2 is involved. In the ultrarelativistic limit, where energy loss by ionization can be neglected, the energy loss is given by the radiation length

$$\frac{dE}{E} = -\frac{dx}{X_0}$$

such that the mean energy $\langle E \rangle$ of an electron with initial energy E_0 after having traversed a mass thickness X is

$$\langle E \rangle = E_0 \, e^{-X/X_0} \tag{1.48}$$

Thus the meaning of the radiation length X_0 is that of a layer thickness which reduces the mean energy of an electron beam by a factor e.

On the other hand, at low velocities energy loss by ionization is the

Fig. 1.4. Mass absorption coefficient for pair production, μ_p/ρ, for some materials.

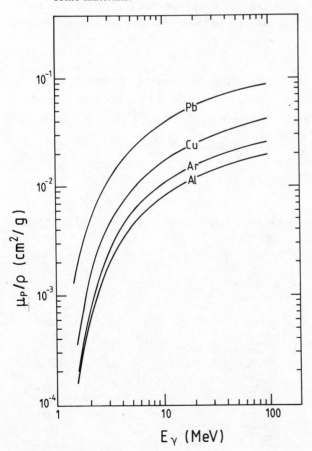

dominant process, and the ratio of energy losses is given approximately by

$$R = \left(\frac{dE}{dx}\right)_{Brem} \bigg/ \left(\frac{dE}{dx}\right)_{ion} \sim \frac{ZE}{580 \text{ MeV}} \tag{1.49}$$

The energy E_c for which the two terms become equal, or $R = 1$, is called the critical energy:

$$E_c \sim \frac{580 \text{ MeV}}{Z} \tag{1.50}$$

More precise values for E_c can be found in table 5.

1.3   Electrons and ions in gases
1.3.1   *Mobility of ions*

If a cloud of ions in a gas is subject to an electric field of strength E, their centre of gravity moves with a constant velocity v_D^+ in the direction of the electric field lines. This average velocity v_D^+ is called the drift velocity. According to experimental results, it depends linearly on the ratio of electric field strength E and gas pressure p. The mobility μ^+ of ions is defined by

$$v_D^+ = \mu^+ E \frac{p_0}{p} \tag{1.51}$$

where $p_0 = 760$ Torr is the standard pressure. Measured values for mobilities of ions in their own gas or in an alien gas are contained in table 6. For a mixture of n different gases, the mobility μ_i^+ of the ion belonging to the gas numbered i is given by

$$\frac{1}{\mu_i^+} = \sum_{k=1}^{n} \frac{c_k}{\mu_{ik}^+} \quad \longleftarrow \text{ concentration of the } k \text{ gas} \tag{1.52}$$

Table 5. *Radiation length X_0, critical energy E_c and hadronic absorption length λ_{had} for some materials*

Material	X_0 (g/cm^2)	E_c (MeV)	λ_{had} (g/cm^2)
H_2	63	340	52.4
Al	24	47	106.4
Ar	20	35	119.7
Fe	13.8	24	131.9
Pb	6.3	6.9	193.7
Lead glass SF 5	9.6	~ 11.8	
Plexiglas	40.5	80	83.6
H_2O	36	93	84.9
NaI(Tl)	9.5	12.5	152.0
$Bi_4Ge_3O_{12}$	8.0	10.5	164

where c_k is the concentration by volume of the gas numbered k and μ_{ik}^+ is the mobility of the ion of kind i in the gas numbered k. If several kinds of ions are present, then those with higher ionization potential are neutralized after 10^2–10^3 collisions by removing electrons from atoms with lower ionization potentials.

1.3.2 *Diffusion of ions in a field-free gas*

According to the equipartition law the average thermic energy of a gas molecule with three degrees of freedom is $\varepsilon_T \sim (3/2)kT$. At a temperature of 273 K, $\varepsilon_T \approx 0.035$ eV. The distribution of kinetic energies ε at a temperature T is

$$F(\varepsilon)=c\sqrt{\varepsilon}\,\exp-\frac{\varepsilon}{kT} \qquad (1.53)$$

A charge distribution, which at $t=0$ is localized at a point $(0,0,0)$, is diffused by multiple scattering into the surrounding volume, whereby a dispersing gaussian density distribution around the origin forms. The diffusion coefficient D is then defined by the following functional form of the differential density distribution dN/N of charges as a function of *one* spatial co-ordinate x at time t:

$$\frac{dN}{N}=\frac{1}{\sqrt{(4\pi Dt)}}\,\exp-(x^2/4Dt)\,dx \qquad (1.54)$$

In other words, the standard deviation of this distribution in one co-ordinate is $\sigma_x=\sqrt{(2Dt)}$. The diffusion coefficient becomes larger with

Table 6. *Measured mobilities of ions in their own gas or in heterogeneous gases* [SA77]

Gas	Ion	Mobility μ^+ (cm^2/V/s)
He	He$^+$	10.2
Ar	Ar$^+$	1.7
H$_2$O	H$_2$O$^+$	0.7
Ar	(OCH$_3$)$_2$CH$_2^+$	1.51
IsoC$_4$H$_{10}$	(OCH$_3$)$_2$CH$_2^+$	0.55
(OCH$_3$)$_2$CH$_2$	(OCH$_3$)$_2$CH$_2^+$	0.26
Ar	IsoC$_4$H$_{10}^+$	1.56
IsoC$_4$H$_{10}$	IsoC$_4$H$_{10}^+$	0.61
Ar	CH$_4^+$	1.87
CH$_4$	CH$_4^+$	2.26
Ar	CO$_2^+$	1.72
CO$_2$	CO$_2^+$	1.09

increasing thermic velocity u of the charged particles. Since $u = \sqrt{(3kT/m)}$, D decreases with increasing mass of the particles. The mean free path during the diffusion process, λ, is given by the collision cross-section $\sigma(\varepsilon)$, which in general depends on the kinetic energy of the charged particles:

$$\lambda(\varepsilon) = \frac{1}{N\sigma(\varepsilon)} \tag{1.55}$$

Here $N = N_0 \rho / A$ is the number of molecules per volume; A is the molar mass; ρ is the density of the gas; and N_0 is Avogadro's constant. For gases at standard conditions (standard temperature and pressure: STP), $N = 2.69 \times 10^9$ molecules/cm^3.

For some molecules and atoms, the diffusion coefficients D^+ are listed in table 7 together with the mean free path λ and the mean thermic velocity u.

The mean free path of electrons, λ_e, is considerably larger than the one for ions. This can be inferred from a simple model in which the particles are treated as solid uncharged spheres. Let a_1 be the radius of an incident particle with velocity $\mathbf{v}_1$, and a_2 the radius of a second kind of particle with velocity $\mathbf{v}_2$, and N the particle density per volume. The relative velocity of the two particles is $\mathbf{v}_1 - \mathbf{v}_2$, and the mean square of this velocity is

$$\langle |\mathbf{v}_1 - \mathbf{v}_2|^2 \rangle = \langle v_1^2 - 2\mathbf{v}_1 \cdot \mathbf{v}_2 + v_2^2 \rangle = \langle v_1^2 + v_2^2 \rangle$$

The mean free path of the first particle is then $\lambda_1 = v_1 \langle t \rangle$, where $\langle t \rangle$ is the mean time between collisions. This time is given by

$$\langle t \rangle = \frac{1}{\pi(a_1^2 + a_2^2)N} \frac{1}{\sqrt{(\langle v_1^2 \rangle + \langle v_2^2 \rangle)}}$$

If the two particles are both atoms or ions of the gas, we have approximately $a_1 = a_2$ and $\langle v_1^2 \rangle = \langle v_2^2 \rangle$, and the mean free path of ions is

$$\lambda_{\text{ion}} = \frac{1}{N\pi a_2^2 4\sqrt{2}}$$

If, on the other hand, the first particle is an electron, then $a_1 \ll a_2$ and $v_1 \gg v_2$,

Table 7. *Thermic velocity u, diffusion coefficient D^+, mobility μ^+ and mean free path λ of ions in their own gas at normal conditions*

Gas	Mass number	u (cm/s)	D^+ (cm^2/s)	μ^+ (cm^2/s/V)	λ (10^{-5} cm)
H_2	2.02	1.8×10^5	0.34	13.0	1.8
He	4.00	1.3×10^5	0.26	10.2	2.8
Ar	39.95	0.41×10^5	0.04	1.7	1.0
O_2	32.00	0.46×10^5	0.06	2.2	1.0
H_2O	18.02	0.61×10^5	0.02	0.7	1.0

such that

$$\lambda_e = \frac{1}{N\pi a_2^2}$$

The two mean free paths are therefore related:

$$\lambda_e = 4\sqrt{2}\,\lambda_{ion} = 5.66\,\lambda_{ion}$$

which is approximately fulfilled for most gases.

1.3.3 Recombination and electron capture
Recombination

The ions and electrons produced during the primary ionization process by a fast charged particle can be neutralized before they are detected. In this process, positive ions recombine with negative ions or with electrons. The decrease of the density of positive ions, n^+, with time can be described by the relation $-dn^+/dt = \alpha n^+ n^-$, where n^- is the density of the negatively charged particles and α is called the 'recombination coefficient'. In unfavourable cases, like O_2 and CO_2, α can reach a value of 10^{-6} cm^3/s for recombination with negative ions, and values up to 10^{-7} cm^3/s for recombination with electrons.

Electron capture

Gas molecules with several atoms are able to accumulate electrons of low (eV) energy. The probability p_a for such an accumulation to happen during one collision is negligibly small for noble gases and for N_2, H_2 and CH_4, but not for electronegative gases like O_2, Cl_2^-, NH_3 and H_2O. This probability for a gas without external electric field is given in table 8 for some cases. If one calculates from the mean thermic electron velocity $u_e \approx \sqrt{(3kT/m_e)}$ and from the mean free path of electrons, λ_e, the number of collisions per time, n_s, then the expression $t_a = 1/(p_a n_s)$ gives the mean time for electron capture.

Table 8. *Probability for electron attachment p_a, collisions per second n_s and mean time for attachment t_a at normal conditions without electric field*

Gas	p_a	n_s (s^{-1})	t_a (ns)
CO_2	6.2×10^{-9}	2.2×10^{11}	7.1×10^5
O_2	2.5×10^{-5}	2.1×10^{11}	190
H_2O	2.5×10^{-5}	2.8×10^{11}	140
Cl	4.8×10^{-4}	4.5×10^{11}	5.0

For strongly electronegative gases this time, at normal conditions, can be as small as 5 ns.

If an electric field is applied, the kinetic energy ε of electrons increases. The probability p_a for electron capture then varies with the electron energy, as shown in fig. 1.5 for O_2. For other gases p_a can be found in the literature [BR 59, LO 61]. If a counting gas contains a fraction f of a contaminating electronegative gas, then the number of collisions of electrons with the molecules of the contaminant is $n'_s = f u_e / \lambda_e$, and the mean free path of electrons relative to electron capture by the electronegative gas is

$$\lambda_a = v_D / (p_a n'_s)$$

where v_D is the drift velocity of electrons. For an admixture of 1 % oxygen in argon at an electric field strength of 1 kV/cm, $\lambda_a \sim 5$ cm. This means that for large drift chambers this effect cannot be neglected, and oxygen filters in the gas cleaning system must be used.

1.3.4   *Drift of electrons in electric fields*

Electrons can gain much more energy in an electric field between two collisions in a gas because their mean free path is longer than that for ions. In particular, the wavelength of electrons in the range of kinetic energies around 1 eV corresponds to two diameters of bound electron orbits in noble gases. This, via a quantum-mechanical effect, makes atoms nearly

Fig. 1.5. Probability for electron capture per collision in O_2 as a function of electron energy ε [BR 59].

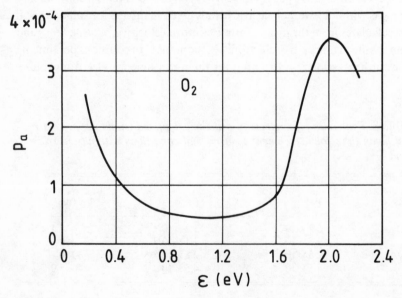

transparent to electrons, i.e. leads to a minimum in the collision cross-section σ as a function of ε ('Ramsauer effect'). Fig. 1.6 shows the dependence of σ on ε for argon in the energy range from 0.1 to 10 eV.

Consider now a swarm of electrons of thermic velocity $u = \sqrt{(2\varepsilon/m)}$, localized at a point P at time $t = 0$. The electrons move away from P in all directions isotropically. The positions at which the first collisions of these electrons with the atoms of the gas happen will lie, on average, on the surface of a sphere with a radius of one mean free path λ_e. If in addition a homogeneous electric field with strength $\mathbf{E} = (0, 0, E)$ in the z-direction is switched on, the electrons will, under the influence of the acceleration $\mathbf{a} = q\mathbf{E}/m$, follow parabolic instead of radial orbits. The penetration points D of these orbits on the surface of the sphere, under the influence of the electric field, are shifted along the surface by $\frac{1}{2}at^2 \sin \theta$, where θ is the angle between the direction of the electric field and the velocity vector of the electron at time $t = 0$ at point P. The z-component of this shift of point D is

$$\delta z = \frac{1}{2} \frac{qE}{m} t^2 \sin^2 \theta$$

Fig. 1.6. Collision cross-section for electrons in argon as a function of electron energy ε [BR 59].

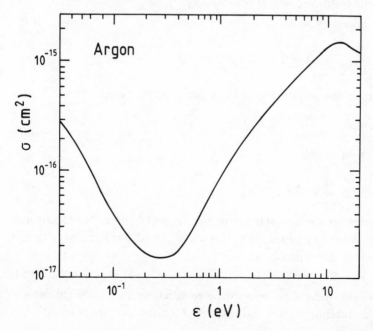

and averaging over $\cos \theta$ gives the mean displacement

$$\langle \delta z \rangle = \frac{1}{3} \frac{qE}{m} \langle t^2 \rangle$$

Assuming the collision cross-section σ and the mean free path λ_e to be independent of the velocity u, the average travelling time over the free path s is

$$\langle t^2 \rangle = \frac{\langle s^2 \rangle}{u^2} = \frac{2\lambda_e^2}{u^2}$$

This leads to the value of the drift velocity

$$v_D = \frac{\langle \delta_z \rangle}{\langle t \rangle} = \frac{2}{3} \frac{qE}{m} \frac{\lambda_e}{u} \tag{1.56}$$

If the velocities u are distributed according to a maxwellian distribution (eq. (1.53)) with a mean square velocity $v = \sqrt{\langle u^2 \rangle}$, this reads

$$v_D = \frac{2}{3} \frac{qE}{m} \frac{1.38 \, \lambda_e}{v} = 0.92 \frac{qE}{m} \frac{\lambda_e}{v} \tag{1.57}$$

The mean time between collisions is

$$\tau = \langle \lambda_e / u \rangle \tag{1.58}$$

In order that electrons adjust to a constant drift velocity, the energy gained by acceleration in the electric field has to be compensated by the energy lost in collisions with atoms. If we call $\Delta(\varepsilon)$ the fraction of the electron energy ε lost in one collision, then, approximately,

$$qE(v_D\tau) = \Delta(\varepsilon)\varepsilon \tag{1.59}$$

or

$$qEv_D = \frac{\Delta(\varepsilon)\varepsilon u}{\lambda_e}$$

Assuming a monochromatic velocity distribution, we get

$$qEv_D \sim \frac{1}{2} \frac{\Delta(\varepsilon)mu^3}{\lambda_e}$$

and using eq. (1.56)

$$v_D \sim \sqrt{\left[\sqrt{(\Delta/2)} \frac{qE}{m} \lambda_e \right]} \tag{1.60}$$

Using simple power laws of the form $\Delta(\varepsilon) \sim \varepsilon^m$ and $\lambda_e(\varepsilon) \sim \varepsilon^{-n}$ as approximations for the energy dependence of Δ and λ_e, the dependence of v_D on the field strength E comes out as

$$v_D \sim E^{(m+1)/(m+2n+1)} \tag{1.61}$$

For low field strength, i.e. below the Ramsauer minimum, one obtains $n \approx -1$ for argon from fig. 1.6, and with $m > 1$ a rapid increase of v_D with E is obtained. At electron energies above the Ramsauer minimum, $n \approx +1$, and the increase of v_D with E is expected to be much slower. Qualitatively, such behaviour is observed in argon and other noble gases.

If, on the other hand, the gas in which the electrons drift is a molecular one (e.g. CO_2, CH_4, Iso-C_4H_{10}), then inelastic collisions contribute significantly to the total cross-section. In CO_2, e.g., molecular oscillations can be excited in the energy range from 0.1 to 1 eV. The fractional energy $\Delta(\varepsilon)$ transferred from the electron to the molecule becomes very large in such inelastic collisions, but decreases again above the maximum excitation energy ε_{max}, approximately as

$$\Delta(\varepsilon) \sim \frac{\varepsilon_{max}}{\varepsilon} \tag{1.62}$$

For $\varepsilon > \varepsilon_{max}$, the exponent $m \sim -1$, and eq. (1.61) gives a drift velocity independent of the electric field strength. If $\Delta(\varepsilon)$ decreases with ε at an even greater rate, then $m < -1$, and v_D decreases with increasing E.

The results of this simplified consideration agree qualitatively with the results of a detailed calculation [PA 75] for which eq. (1.56) is replaced by

$$v_D = \frac{qE}{m} \left(\frac{2}{3} \left\langle \frac{\lambda_e(u)}{u} \right\rangle + \frac{1}{3} \left\langle \frac{d\lambda_e(u)}{du} \right\rangle \right) \tag{1.63}$$

With these calculations it is possible to reproduce quantitatively the observed dependence of drift velocities on the electric field strength E. Fig. 1.7 shows experimental results for one-component gases, fig. 1.8 for argon–isobutane mixtures and fig. 1.9 for argon–methane mixtures. A characteristic feature of the 'fast' gas methane is the decrease of v_D at high field strength ($E > 1$ kV/cm), which is present also in mixtures of argon with methane.

1.3.5 Drift of electrons in electric and magnetic fields

A particle with charge q and velocity $\mathbf{v}$ is subject to the Coulomb force $q\mathbf{E}$ in an electric field $\mathbf{E}$ and to the Lorentz force $q\mathbf{v} \times \mathbf{B}$ in a magnetic field $\mathbf{B}$. In the magnetic field, the particle energy is not changed, and the particles follow a circular or helical orbit with angular velocity

$$\omega = -\frac{q\mathbf{B}}{m} \tag{1.64}$$

The modulus $\omega = |\omega|$ is the cyclotron frequency and has the value

$$\frac{\omega}{B} = 17.6 \text{ MHz/G} \tag{1.65}$$

for electrons.

If electric and magnetic fields are present at the same time, the orbit is a helix, and the movement can be decomposed into a circular one with angular velocity ω and a translation with velocity $\mathbf{v_D}$ in the following way:

$$\mathbf{v} = \mathbf{v_D} + \omega \times \mathbf{r_b}$$

with

$$\mathbf{v_D} = \mathbf{E} \times \frac{\mathbf{B}}{B^2} + \mathbf{v}_{\parallel} \tag{1.66}$$

and

$$m\dot{\mathbf{v}}_{\parallel} = q\mathbf{E}_{\parallel}$$

where $\mathbf{v}_{\parallel}$ and $\mathbf{E}_{\parallel}$ are the components of the vectors parallel to $\mathbf{B}$, and $\mathbf{r_b}$ designates the position of the particle on a plane perpendicular to $\mathbf{v_D}$. If the particle now moves in a gas-filled volume, the collisions with the gas molecules can be described by an additional stochastic force $m\mathbf{A}(t)$ which varies with time. The equation of movement is then

$$m\dot{\mathbf{v}} = q(\mathbf{E} + \mathbf{v} \times \mathbf{B}) + m\mathbf{A}(t) \tag{1.67}$$

Fig. 1.7. Drift velocity of electrons in gases at normal conditions [EN 53, FU 58, BR 59].

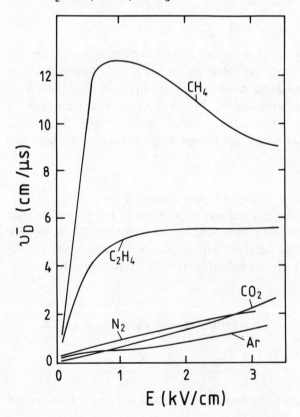

Fig. 1.8. Drift velocity of electrons in argon–isobutane mixtures [BR 74].

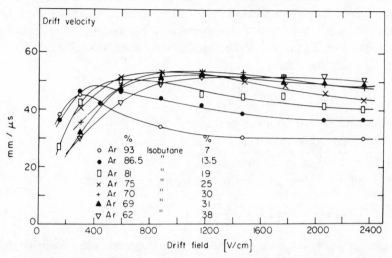

Fig. 1.9. Drift velocity of electrons in argon–methane mixtures [TI 83]. The dashed curve is the result of a calculation for the 80/20 mixture [SC 78].

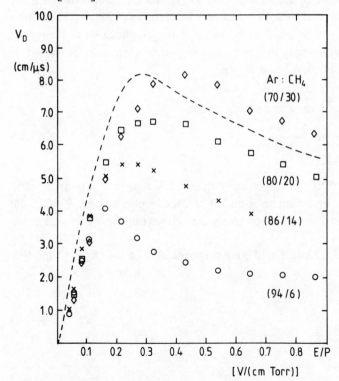

This equation, called the 'Langevin equation', can be averaged in time. Since the solution is known to be a translation with constant drift velocity, the average acceleration has to vanish. This means that averaged over time the stochastic acceleration compensates the translational acceleration $\langle \mathbf{A}(t) \rangle = -\mathbf{v}_D/\tau$, where τ is the mean time between two collisions.

Eq. (1.67) then reads:

$$\dot{\mathbf{v}}_D = \frac{q\mathbf{E}}{m} + \left(\mathbf{v}_D \times \frac{q\mathbf{B}}{m}\right) - \frac{\mathbf{v}_D}{\tau}$$

or, because $\dot{\mathbf{v}}_D = 0$ for a constant electric field:

$$\frac{\mathbf{v}_D}{\tau} + \left(\frac{q\mathbf{B}}{m} \times \mathbf{v}_D\right) = \frac{q\mathbf{E}}{m} \tag{1.68}$$

The following expression for $\mathbf{v}_D$ is a solution of eq. (1.68):

$$\mathbf{v}_D = \frac{\mu}{1+\omega^2\tau^2}\left(\mathbf{E} + \frac{\mathbf{E} \times \mathbf{B}}{B}\,\omega\tau + \frac{(\mathbf{E} \cdot \mathbf{B}) \cdot \mathbf{B}}{B^2}\,\omega^2\tau^2\right) \tag{1.69}$$

where $\mu = q\tau/m$ is the electron mobility. The translational velocity or drift velocity $\mathbf{v}_D$ in the presence of electric *and* magnetic fields consists of three components: one parallel to $\mathbf{E}$, one parallel to $\mathbf{B}$, and one perpendicular to the plane spanned by $\mathbf{E}$ and $\mathbf{B}$. For $\omega\tau = 0$, $\mathbf{v}_D$ follows the direction of $\mathbf{E}$, for $\omega\tau \gg 1$, $\mathbf{v}_D$ is parallel to $\mathbf{B}$. For arbitrary finite values of $\omega\tau$ the penetration point of the direction given by $\mathbf{v}_D$ with a plane perpendicular to $\mathbf{B}$ lies on a semi-circle connecting the penetration points of $\mathbf{E}$ and $\mathbf{B}$. For $\mathbf{E} = (E_x, 0, E_z)$, $\mathbf{B} = (0, 0, B_z)$ and $E_x \ll E_z$:

$$v_x = \mu E_x \frac{1}{1+\omega^2\tau^2}$$

$$v_y = -\mu E_x \frac{\omega\tau}{1+\omega^2\tau^2} \tag{1.70}$$

$$v_z = \mu E_z$$

These relations for the direction of $\mathbf{v}_D$ as a function of $\omega\tau$ are depicted in fig. 1.10. In this case where $\mathbf{E}$ and $\mathbf{B}$ are nearly parallel and $E_z \gg E_x$, the modulus of the drift velocity is nearly the same as the one without magnetic field.

If, on the other hand, $\mathbf{E}$ and $\mathbf{B}$ are perpendicular, i.e. $\mathbf{E} = (E_x, 0, 0)$ and $\mathbf{B} = (0, 0, B_z)$, then

$$v_x = \mu E_x \frac{1}{1+\omega^2\tau^2}$$

$$v_y = -\mu E_x \frac{\omega\tau}{1+\omega^2\tau^2} \tag{1.71}$$

$$v_z = 0$$

In this case, the drift velocity depends on the magnetic field strength:

$$v_D = \sqrt{(v_x^2 + v_y^2)} = \mu E_x \frac{1}{\sqrt{(1 + \omega^2 \tau^2)}} \qquad (1.72)$$

and the angle between v_D and E, the 'Lorentz angle' α, is given by

$$\tan \alpha = \omega \tau \qquad (1.73)$$

These last two relations describe the experimental data well, as shown in fig. 1.11.

1.3.6 *Diffusion of electrons in electric and magnetic fields*

The magnetic field B not only influences the drift direction and the drift velocity, but also diminishes the diffusion coefficient of electrons in a direction transverse to the B field direction [AL 56]. If $B = (0, 0, B_z)$ and the diffusion coefficient in the field-free gas is D, then in the presence of the magnetic field the diffusion coefficients in the three cartesian directions are

$$D_z = D$$
$$D_x = D_y = \frac{D}{(1 + \omega^2 \tau^2)} \qquad (1.74)$$

The coefficient for diffusion transverse to the magnetic field is reduced considerably if for an electron velocity u the bending radius in the magnetic field, u/ω, is small compared with the mean free path $\lambda = u\tau$, i.e. if $\omega\tau \gg 1$.

The diffusion coefficient is also not isotropical if only an electric field is present [PA 68]. The coefficient defined in section 1.3.2 remains valid for diffusion transverse to the electric field E, while the coefficient D_L for

Fig. 1.10. Direction of the drift velocity vector v_D in the presence of a magnetic field B along the z-direction and an electric field E. The circle connects the points of penetration of B and E in the (x, y)-plane. The penetration point of v_D lies on this circle.

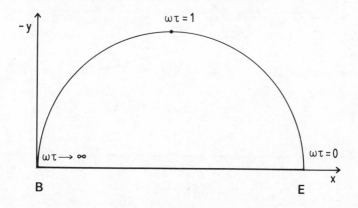

diffusion parallel to $\mathbf{E}$ is not equal to D for some gases (argon, for example). A measurement of the anisotropy D_L/D for a mixture of argon and methane (9:1) with 28% isobutane is shown in fig. 1.12. For most other gases, $D_L/D \approx 1$.

In a track detector one tries to minimize the diffusion transverse to the drift direction of electrons in order to obtain a 'sharp' image of the primary ionization at the end of the drift path. This optimization leads to different results depending on whether a magnetic field parallel to $\mathbf{E}$ is present or not. Let L be the drift path, v_D the drift velocity, u the mean velocity of electrons and λ their mean free path. For $B = 0$ we obtain from section 1.3.2 $\sigma = \sqrt{(2Dt)}$ and together with $t = L/v_D$ and $D = u\lambda/3$:

$$\sigma = \sqrt{\left(\frac{2L}{3v_D}\right)} \sqrt{u\lambda} \tag{1.75}$$

The transverse diffusion in this case is minimized by choosing a gas with a very small mean free path, e.g. CO_2. Fig. 1.13 shows the variance σ of transverse diffusion for a drift length of $L = 15$ cm in mixtures of argon with methane and CO_2, as a function of the drift field strength E.

If, on the other hand, an additional magnetic field is switched on, the diffusion transverse to this field is reduced by a factor $1/(1 + \omega^2\tau^2)$. The situation is transparent if the magnetic field is parallel to the electric one.

Fig. 1.11. Electron drift velocity v_D and Lorentz angle α as a function of the magnetic flux density B for crossed electric and magnetic fields [BR 75].

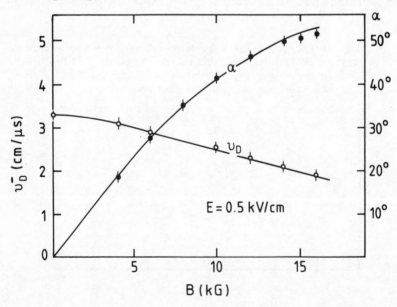

Then with $\tau = \lambda/u$

$$\sigma(B) = \sqrt{\left(\frac{2L}{3v_D}\right)} \bigg/ \sqrt{\left(\frac{u\lambda}{1+\omega^2\lambda^2/u^2}\right)} \tag{1.76}$$

For $\omega\tau \gg 1$, i.e. for strong magnetic fields, $\sigma(B)$ becomes minimal for the largest possible mean free path: the larger the mean free path, the more time is available for the magnetic field to keep the electrons on their helical orbits. Measurements corresponding to this situation are shown in fig. 1.14 for a magnetic field with $B = 20.4$ kG parallel to the electric drift field. While for argon–CO_2 mixture the reduction of the transverse diffusion is only 35% compared with the case without magnetic field, for the argon–methane mixture reductions by one order of magnitude are observed. Since for $\omega\tau \gg 1$ the ratio $\sigma/\sigma(B) = \omega\tau$, the two measurements at $B = 0$ and at $B = 20.4$ kG can be used to extract the product $\omega\tau$, and, with the help of eq. (1.65) for ω, to obtain τ. This is shown in fig. 1.15. The mean time between two collisions for this gas mixture is of the order 10^{-11} s. The maximum of $\omega\tau$ corresponds to a minimum collision cross-section due to the Ramsauer effect (section 1.3.4). The electron energy at which this happens, the

Fig. 1.12. Anisotropy of diffusion coefficients parallel (D_L) and perpendicular to the electric field as a function of the reduced electric field E/p (units V/(cm Torr)) or E/N (units $Td = 10^{-17}$ V cm^2). Measured points from [SC 80], calculated curves from [PA 68] and [RO 72].

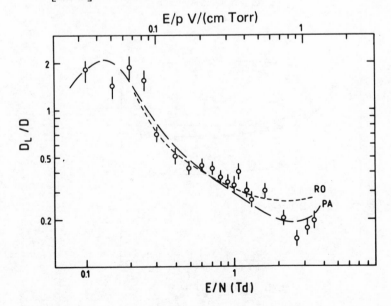

'characteristic energy' ε_K, is defined as:

$$\varepsilon_K = \frac{D}{\mu} \tag{1.77}$$

It can be calculated from the value E_0 of the electric field at this maximum of $\omega\tau$ that

$$\frac{\sigma^2}{2L} = \frac{D}{v_D} = \frac{\varepsilon_K \mu}{v_D} = \frac{\varepsilon_K}{E_0} \tag{1.78}$$

For a mixture with 10% methane, one reads from fig. 1.15 that $E_0 = 0.1$ V/(Torr cm) which gives $\varepsilon_K = 0.15$ eV, in agreement with the energy at the Ramsauer minimum for argon in fig. 1.6.

1.4 Quantities characterizing detectors

In general, a particle will be detected in a measuring instrument by the charge Q liberated during the passage of the particle at $t = 0$ in one of the

Fig. 1.13. Standard deviation of the transverse diffusion broadening of a point-like electron cloud drifting along an electric field **E** in an argon–methane mixture [PE 76]. Drift length is 15 cm.

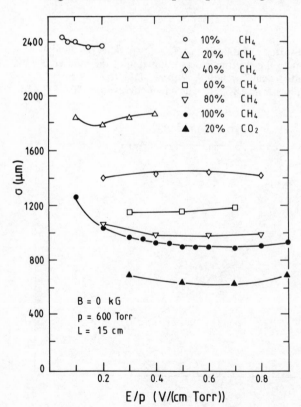

processes described in section 1.2 or by light quanta. This charge is guided towards an electrode by means of electric, and sometimes magnetic, fields and collected there. The collection time t_c can vary between a few nanoseconds in semiconductor detectors or photomultipliers and a few milliseconds in ionization chambers. The instrument will therefore produce a current flowing between time $t = 0$ and $t = t_c$, and the integral over this current is Q.

If the information about the precise time of passage of the particle is irrelevant for the purpose of the measurement, the simplest way of recording the detector output is a measurement of the average dc current delivered by the detector. This can serve as a measurement of the rate of incident particles if the amount of charge liberated by each individual particle is the same. This current mode of operation is used for radiation dosimetry.

In most applications, however, the timing information is important, and

Fig. 1.14. Same as fig. 1.13, but with a magnetic field **B** parallel to **E**; $|\mathbf{B}| = 20.4$ kG [PE 76].

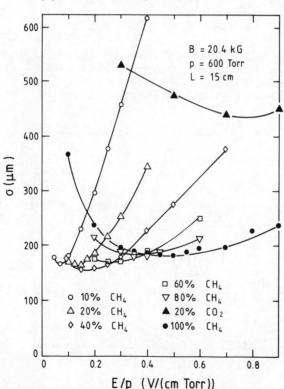

Fig. 1.15. Values for $\omega\tau$ deduced from figs. 1.13 and 1.14 (see text).

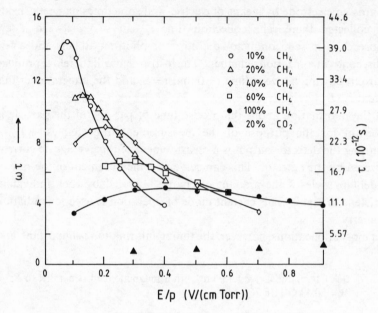

Fig. 1.16. Pulse height distribution of scintillation light produced by α-particles from an ^{241}Am source embedded in a plastic scintillator. P_0: peak pulse height; FWHM: full width at half maximum; $\sigma =$ FWHM/2.36: standard deviation. The energy resolution is $\sigma/P_0 \sim 12\%$.

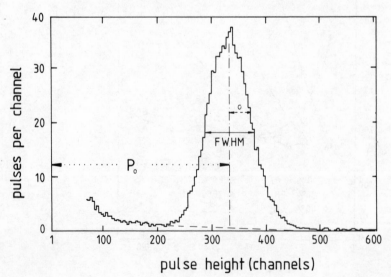

therefore the detector output for each individual particle is recorded. In this *pulse mode* of operation the output current of the detector is transformed to a voltage signal by a preamplifier. The time structure of this signal is then determined by the input impedance of this circuit, which can be realized by an input resistor R_i parallel to an input capacitor C_i. If the time constant $\tau = R_i C_i$ is small compared with the collection time t_c, the signal follows the detector output closely.

If, however, $RC \gg t_c$, the voltage U on the input capacitor C_i rises until the charge Q is completely collected, and at that time $t = t_c$ the maximum $U_{max} = Q/C_i$ is reached. Therefore, in this case the risetime of the voltage pulse is determined by the charge collection time of the detector, and its decay time by the time constant $R_i C_i$ for the circuit.

For some detectors (e.g. semiconductor detectors), the capacitance C_i cannot be kept constant. In this case a charge sensitive amplifier is used which removes the dependence of the output voltage on C_i. This is an inverting amplifier with a feedback loop through a capacitance C_f. The amplification A is large compared with the ratio $(C_f + C_i)/C_f$. Then the output voltage is approximately proportional to the input charge Q:

$$V_{out} = -A \frac{Q}{C_f(A+1)+C_i} \sim -\frac{Q}{C_f} \qquad (1.79)$$

and this signal is independent of the input capacitance.

Thus, for each detected particle, the detector and associated electronics deliver a voltage pulse proportional to the collected charge. This information is sometimes converted from an analog into a digital form by analog-to-digital converters (ADC). The differential distribution dN/dP of this pulse height information, i.e. the number of pulses per interval of P, is called a *pulse height spectrum*.

One characteristic property of a detector is its *resolution* for measuring a certain quantity Z. Let z be the response of the detector, then the resolution is defined as the standard deviation σ_z or the full width at half maximum (FWHM) Δz of the distribution $D(z)$ in the measured quantity z for a monochromatic input distribution, $\delta(Z - \langle Z \rangle)$. The mean value of the measured quantity is $\langle z \rangle = \int zD(z)\,dz$, the variance is $\sigma_z^2 = \int (z - \langle z \rangle)^2\,dz$, and the standard deviation σ_z is the square root of the variance. As an example consider the differential pulse height distribution in fig. 1.16. Here an α-active source of ^{241}Am is distributed homogenously in an organic scintillator. It emits mono-energetic α-particles. The scintillation light is registered in a photomultiplier tube, and the output pulse is converted into channels of an analog-to-digital converter in such a way that the channel number (abscissa in fig. 1.16) is proportional to the original number of photoelectrons. In this figure, the meaning of Δz (FWHM) and of $\sigma_z = \Delta z/2.36$ is depicted.

This relation $\sigma_z = \Delta z/2.36$ holds if the distribution function $D(z)$ for a monochromatic input is a gaussian distribution. If $D(z)$ is a box distribution with width Δz, the standard deviation amounts to $\sigma_z = \Delta z/\sqrt{12}$.

The values Δz and σ_z are absolute quantities, measured in units of z. If the relation between input quantity Z and response z is known, the corresponding absolute quantities ΔZ and σ_Z are known in units of Z. The relative resolution is defined as the dimensionless ratio $\sigma_z/\langle z \rangle$ between standard deviation and mean of the measured quantity.

If the only source of fluctuations in the signal is the statistical fluctuation of the number N of primary charge carriers and if their formation follows Poisson statistics, then for $N > 20$ the response function is approximately a gaussian distribution, and the relative resolution is given by the standard deviation of the gaussian distribution:

$$\frac{\sigma_z}{\langle z \rangle} = \frac{\sqrt{N}}{N} = \frac{1}{\sqrt{N}} \tag{1.80}$$

In principle, this value should be a lower limit to the resolution. Contrary to this expectation, it has been found in several types of detectors that the resolution can be smaller than this value by a factor up to four. This observation indicates that the assumption about the validity of Poisson statistics was false, and that there is a correlation between the processes giving rise to the formation of individual charge carriers. This phenomenon is called the Fano effect, and one introduces a Fano factor F:

$$F = \left(\frac{\text{observed resolution}}{\text{resolution expected from Poisson statistics}} \right)^2 \tag{1.81}$$

The Fano factor ranges from ~ 0.06 for semiconductor detectors to unity for scintillation counters. Intermediate values around 0.17 apply for noble gases like neon or argon.

The relation between the measured mean $\langle z \rangle$ and the value Z of the original quantity of the incident particles has to be established by a calibration of the detector in a monochromatic beam of particles. If the relation is of the type

$$\langle z \rangle = cZ \tag{1.82}$$

with a constant value c, the response of the detector is called 'linear'. If c varies with Z, the relative variation $Z/c(dc/dZ)$ is called non-linearity.

Another important aspect of a detector is the *detection efficiency*, i.e. the probability of recording a pulse from a particle emitted in an elementary reaction. This absolute efficiency is composed of two factors: the *solid angle* Ω subtended by the detector as seen from the source or interaction point, and the intrinsic efficiency ε_i of the detector. The solid angle can be calculated from the geometry of the experiment. For a point source it is

$$\Omega = \int_D (-\mathrm{d} \cos \theta) \, \mathrm{d}\varphi \qquad (1.83)$$

If, for example, the detector surface is a circle with radius r at a distance R from the source, then

$$\Omega = 2\pi(1 - \cos \alpha) = 2\pi\left(1 - \frac{R}{\sqrt{(r^2 + R^2)}}\right)$$

with $\sin \alpha = r/\sqrt{(r^2 + R^2)}$, and for $r \ll R$ this gives

$$\Omega = \frac{\pi r^2}{R^2} \qquad (1.84)$$

The *intrinsic* efficiency ε_i of the detector is defined as the number of signals recorded divided by the number of particles impinging on the detector surface. It depends on the probability of interaction of the particle, on the efficiency for collecting the charges or light quanta emitted, and on electronic thresholds in the detector.

The intrinsic efficiency can also be reduced if the detector is unable to process an event because it is still busy with the preceding one. This phenomenon occurs at high counting rates and is called *deadtime*. Two types of behaviour can be realized: in a *non-paralysable* detector each recorded signal is followed by a time interval of length τ during which no new true event is accepted. If the rate of true events is called R and the one of recorded events R', then the fraction of time during which the detector is dead is $R'\tau$. The rate of true events lost by this deadtime is $RR'\tau$. Since this is equal to $R - R'$, we obtain the true event rate:

$$R = \frac{R'}{1 - R'\tau} \qquad (1.85)$$

In special cases it is also desirable to block the detector in those cases where the true event happens during the deadtime of the preceding event. This is called the *paralysable* mode of operation. Then the deadtime intervals are of variable length. The rate of recorded events R' in this case is the same as the rate at which time intervals larger than τ occur in the true event sequence. According to Poisson statistics the probability of obtaining an interval larger than τ is $\mathrm{e}^{-R\tau}$, and thus the rate at which this occurs is $R \, \mathrm{e}^{-R\tau}$. Therefore

$$R' = R \, \mathrm{e}^{-R\tau} \qquad (1.86)$$

which cannot be solved explicitly for the true rate R.

For low event rates, i.e. $R \ll 1/\tau$, both eqs (1.85) and (1.86) give approximately the same result

$$R' \approx R(1 - R\tau)$$

or   (1.87)

$$R \approx R'(1 + R'\tau)$$

2

....

Measurement of ionization

2.1 Ionization chambers

In these chambers the ionization produced by a passing particle is measured. They can be operated in a pulse mode or in a current mode. In the first case the pulse induced by each individual incident particle is measured independently; in the second case the average dc current produced by the sequence of incident particles is recorded.

Pulse mode ionization chamber

The simplest form of such a chamber is a parallel-plate capacitor filled with a counting gas, e.g. argon (fig. 2.1). The value of the electric field strength $|\mathbf{E}| = E_z = U_0/d$ must be such that the N positive and N negative charges formed along the path of the passing particle at $z = z_0$ are collected completely on the capacitor plates, but that no secondary ionization by the drifting electrons takes place. The charges moving in the electric field induce a charge on the capacitor electrodes. This charge flows through the resistor R and can be measured as a voltage pulse. If the path of the passing particle is parallel to the electrodes at $z = z_0$, this pulse can be calculated by considering energy conservation when shifting N charges from z_0 to z:

$$\frac{1}{2}CU^2 = \frac{1}{2}CU_0^2 - N\int_{z_0}^{z} qE_z \, \mathrm{d}z \tag{2.1}$$

where C is the capacitance of the anode against ground. From this we obtain

$$\frac{1}{2}C2U_0\,\Delta U = -\frac{NqU_0}{d}(z-z_0)$$

and

$$\Delta U = -\frac{Nq}{Cd}(z-z_0) \tag{2.2}$$

For a constant drift velocity v_D^+ for ions and v_D^- for electrons one obtains

$$\Delta U^+ = -\frac{Ne}{Cd} v_D^+ \, \Delta t^+$$

$$\Delta U^- = -\frac{N(-e)}{Cd} (-v_D^-) \, \Delta t^-$$

(2.3)

The polarity of the two contributions to the pulse is the same because the opposite charges drift in opposite directions. Since the electrons drift much faster than the ions, the pulse increases first due to the electron movement up to a value of $\Delta U = -(Nez_0)/Cd$ (assuming $R = \infty$ for the moment), and then continues to rise slower due to ion movement up to the asymptotic value $\Delta U = -Ne/C$ (see fig. 2.2). The collection time in an argon chamber with $d = 5$ cm at STP and an electric field of 500 V/cm amounts to $\Delta t^- \sim$ 1 μs for electrons, but up to $\Delta t^+ \sim 1$ ms for ions. The voltage pulse is independent of the position z_0 at which the primary ionization occurred

Fig. 2.1. Parallel-plate ionization chamber (schematic).

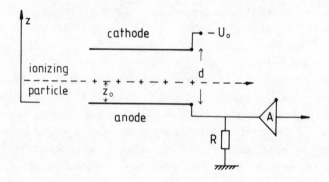

Fig. 2.2. Time development of voltage pulse $\Delta U(t)$ from planar ionization chamber for $R = \infty$.

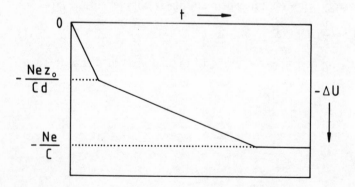

only if the time constant RC exceeds the collection time Δt^+ of positive ions. This long time is impractical for counting individual particles. It is then possible to reduce the effective time constant of the detector by introducing an $R'C'$-coupling in front of the amplifier A in fig. 2.1. If then we choose $\Delta t^- < R'C' \ll \Delta t^+$, the contribution of the ions to the pulse is neglected and the pulse is given by

$$\Delta U = -\frac{Ne}{C}\frac{z_0}{d}$$

The amplitude of this pulse depends on the position z_0 at which the electrons were formed.

If the time constant is chosen to be even smaller, $R'C' \ll \Delta t^-$, then only the first part of the electron-induced pulse is recorded, and the amplitude is nearly independent of the position z_0 according to eq. (2.3). However, the signal here is small compared with the one available with long integration time.

In order to avoid the problem encountered for the case $\Delta t^- < R'C' \ll \Delta t^+$ of having a signal depending on the position of the ionizing track, a way out was found by O. Frisch [FR 44]. It consists of inserting a 'Frisch grid' between the anode and the cathode at an intermediate potential. By collimation, the incident particles are confined to the region between the cathode and the grid. Ions then migrate to the cathode as before. Electrons from the ionization process, however, drift towards the grid, traverse it and continue on their way to the anode. If the pulse is only collected between the grid and the anode, each electron has the same drift time between the grid and the anode and therefore all electrons contribute the same amount to the pulse. The pulse amplitude is independent of position, and the integration time can be made much smaller than the ion collection time, but still such that the total electron contribution is caught: $\Delta t^- < R'C' \ll \Delta t^+$.

If instead of a homogenous electric field, as in the parallel-plate capacitor, a cylindrical field is used, one obtains the arrangement shown in fig. 2.3.

The electric field strength as a function of the radial distance r from the axis of the cylinder is

$$E(r) = \frac{U_0}{r\ln(r_a/r_i)} \tag{2.4}$$

For electrons formed at a distance $r = r_0$ the drift time up to the anode is

$$\Delta t^- = \int_{r_i}^{r_0} \frac{dr}{v_D^-}$$

For values of $E/p \sim 0.1$ V/(Torr cm) the electron drift velocity is proportional to E, $v_D^- = \mu^- E$, such that

$$\Delta t^- \approx \int_{r_i}^{r_0} \frac{dr}{\mu^- E} = \frac{\ln(r_a/r_i)}{\mu^- U_0} \int_{r_i}^{r_0} r \, dr = \frac{\ln(r_a/r_i)}{2\mu^- U_0} (r_0^2 - r_i^2) \quad (2.5)$$

The pulse generated by the movement of the electrons can again be obtained from energy conservation (eq. (2.1)):

$$\Delta U^- = -\frac{Ne}{C} \frac{\ln(r_0/r_i)}{\ln(r_a/r_i)} \quad (2.6)$$

The amplitude of this pulse does not depend linearly on the distance between the ionizing track and the anode, as it was the case for the parallel-plate chamber, but logarithmically. The contribution of the moving positive ions is obtained in the same way:

$$\Delta U^+ = -\frac{Ne}{C} \frac{\ln(r_a/r_0)}{\ln(r_a/r_i)} \quad (2.7)$$

If $r_a \gg r_i$ and if the cylindrical chamber is irradiated homogenously, the contribution of the electron component is the dominating one. As an example, for $r_a/r_i = 10^3$ and $r_0 = r_a/2$ the ratio $\Delta U^+/\Delta U^- = \ln 2/\ln 500 \approx 0.1$.

Current mode ionization chamber
If the resistance R in figs 2.1 and 2.3 is very large ($RC > \Delta t^+$) then a measurement of individual particles entering the counter at a rate of more than 1 kHz is not possible. In this case, an average dc current can be

Fig. 2.3. Cylindrical ionization chamber (schematic).

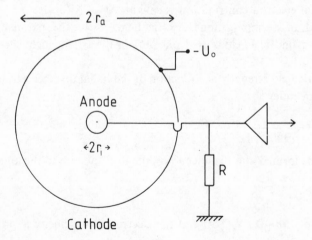

measured:

$$I = -RNe \qquad (2.8)$$

This current can be used to measure the γ-ray exposure and to calibrate the intensity of radioactive sources (α- and β-emitters). For the measurement of γ-ray exposure, usually chambers filled with air are used, since the dose units are defined in air (section 1.1.2). Fig. 2.4 shows a chamber used for measurement of radiation dose, with walls made of tissue-equivalent material (polystyrene mixed with graphite or aluminium).

2.2 Proportional counters

This counter differs from the ionization chamber discussed above by the higher electric fields used. In proportional counters the electrons formed in the primary ionization process enter an electric field of such a strength (10^4–10^5 V/cm) that they can gain, between two collisions, enough kinetic energy to be able to ionize other atoms or molecules of the gas. In a cylindrical electric field (see eq. (2.4)), a primary electron gains kinetic

Fig. 2.4. Ionization chamber used for ion dose measurements. Material is tissue-equivalent synthetic product (after [RO 56]).

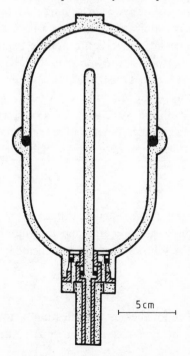

5 cm

energy between collisions at radial distances r_1 and r_2

$$\Delta T_{\text{kin}} = e \int_{r_1}^{r_2} E(r) \, dr$$
$$= eU_0 \frac{\ln(r_2/r_1)}{\ln(r_a/r_i)} \tag{2.9}$$

If ΔT_{kin} exceeds the ionization potential of the counting gas, secondary ionization takes place. A chain of such reactions leads to an avalanche of electrons and ions formed ('Townsend avalanche'). The amount of charge liberated in the process is amplified by a factor A compared with the free charge formed in the primary ionization, Ne. The voltage pulse detected at the electrodes also increases by A:

$$\Delta U = -A \frac{Ne}{C} \tag{2.10}$$

A is called the *gas amplification factor*. There is a region of electric field strength E and gas pressure p in which this amplification factor A is independent of the amount of primary ionization. This means that the pulse measured is *proportional* to the charge from the primary ionization process. This domain of electric field strength is called the *proportional region*, and the gas amplification factor is in the range of 10^4–10^6 in this region.

The high electric field strength required for gas amplification is achieved by using a thin (20–100 μm dia.) wire as the anode in a cylindrical counting tube. According to eq. (2.4), secondary ionization processes will start in the close neighbourhood of the wire once the primary ionization electrons have migrated to this region. The number of electron–ion pairs formed by an electron along a path of 1 cm length is called the first Townsend coefficient α. It can be calculated from fig. 2.5 using the relation $\alpha = \sigma_i N$ where σ_i is the cross-section for ionization by collision and $N = 2.69 \times 10^{19}$ atoms/cm^3, the atomic density of noble gases at STP. If the number of primary electrons at point $x = 0$ is n_0, the number $N(x)$ of electrons present after a path length x is obtained from the relation

$$dN(x) = N(x)\alpha \, dx$$

with the solution $N(x) = n_0 \exp \alpha x$, if α is independent of x.

In general, α will depend on the electron energy and therefore on the electric field strength, such that the solution is more complicated:

$$N = n_0 \exp\left(\int \alpha(x) \, dx \right) \tag{2.11}$$

This relation gives therefore a value for the gas amplification factor, $A = \exp(\int \alpha(x) \, dx)$. This result can also be expressed in terms of the mean free path for collisions of electrons in the gas. Since

$$\lambda = \frac{1}{\alpha} = \frac{1}{N\sigma_i}$$ (2.12)

the relation is

$$A = \exp \int \frac{dx}{\lambda(x)}$$

The dependence of the gas amplification on the voltage U_0 applied to the counter can be calculated approximately if one neglects recombination of ions and electron attachment and assumes that no ionization happens through ultraviolet photons from excited atoms. One then obtains:

$$A \propto \exp\left[k\sqrt{U_0}\left(\sqrt{\frac{U_0}{U_s}} - 1\right)\right]$$ (2.13)

where U_s is the threshold voltage at which the gas amplification sets in, and

Fig. 2.5. Cross-section for ionization by collision for electrons of energy ε in noble gases [AL 69].

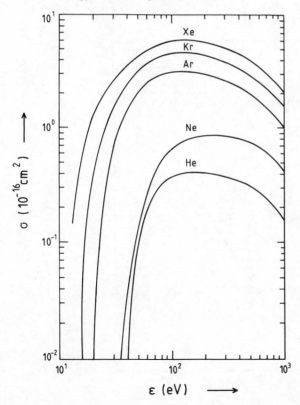

k is a constant. Eq. (2.13) is valid for small gas amplifications ($A \lesssim 10^3$) as shown in fig. 2.6.

This exponential rise of A with U_0 in the proportional region ends when the number of ionization processes due to photons in the ultraviolet wavelength region becomes sizeable. These photons originate from the deexcitation of atoms excited in collisions, and they produce electrons by the photoelectric effect in gas atoms or in the cylindrical cathode. In the avalanche formation, started by n_0 primary electrons, $n_0 A$ electrons are formed; assume that at the same time $(n_0 A)\gamma$ photoelectrons are produced by the ultraviolet transfer process, with a coefficient $\gamma \ll 1$. Then by gas amplification these photoelectrons give rise to $n_0 A^2 \gamma$ electrons. In these avalanches $n_0 A^2 \gamma^2$ photoelectrons are formed, and $n_0 A^3 \gamma^2$ electrons are liberated. Adding the number of electrons from the sequential steps, one obtains:

$$A_\gamma = n_0 A \sum_{n \geq 0} (A\gamma)^n = \frac{n_0 A}{1 - A\gamma} \tag{2.14}$$

Fig. 2.6. Gas amplification factor A in an argon proportional counter as a function of voltage U for two gas pressures [ST 53]; measured values ($+$) and calculated curve.

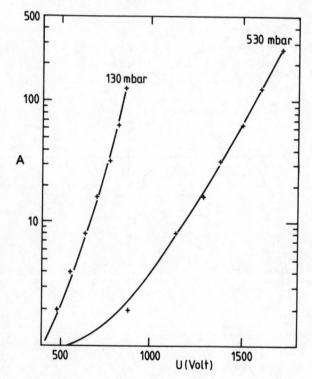

This is the gas amplification factor including the energy transfer by photons. For $A\gamma \to 1$, the expression diverges, and the signal output no longer depends on the primary ionization. This region of operation is called the *Geiger–Mueller region* or trigger region. The border between the region, where still a *limited proportionality* between primary ionization and total liberated charge exists, and the trigger region is given approximately by $\alpha x \sim 20$ or $A \sim 10^8$.

We now study the pulse formation in a proportional counter. For this purpose we use eqs (2.6) and (2.7) which are valid for the cylindrical ionization chamber. The charge carriers responsible for the pulse are no longer the electrons and ions from the primary ionization process but those charges formed in the avalanches near to the anode wire. The radial position r_0 of the origin of these charge carriers is limited to a distance of a few (k) mean free paths from the wire surface: $r_0 = r_i + k\lambda$. Therefore the ratio of contributions of ion movement and electron movement to the pulse is large here (for $\lambda \ll r_i$):

$$R = \frac{\Delta U^+}{\Delta U^-} \sim \frac{\ln(r_a/r_i)}{\ln((r_i + k\lambda)/r_i)} \sim \frac{\ln(r_a/r_i)}{k\lambda/r_i} \qquad (2.15)$$

With $r_a = 20$ mm, $r_i = 0.1$ mm and $k\lambda = 0.02$ mm one obtains for argon at STP $R \approx 25$. In a proportional counter the main contribution to the signal comes from the ions in the avalanche slowly drifting away from the wire and not from the electrons moving at a much faster rate towards the wire. The contribution of the electron component can be increased by reducing the gas pressure, i.e. by increasing the mean free path λ of electrons. The risetime of the electron-induced pulse or the collection time of electrons is according to eq. (2.5):

$$\Delta t^- = \frac{(r_0^2 - r_i^2)\ln(r_a/r_i)}{2\mu^- U_0}$$

For electron mobilities of $\mu^- \sim 10^2 – 10^3$ cm^2/V s and $U_0 = 10^2$ V with the dimensions of the counter used above this time is about $\Delta t^- \sim 10^{-8} – 10^{-9}$ s. For the positive ions we have

$$\Delta t^+ = \frac{\ln(r_a/r_i)}{\mu^+ U_0}\int_{r_0}^{r_a} r\, dr \approx \frac{(r_a^2 - r_i^2)\ln(r_a/r_i)}{2\mu^+ U_0} \qquad (2.16)$$

This collection time is much larger than Δt^- because of the small mobility of ions, $\mu^+ \sim 1$ cm^2/V s, and because the ions have to drift over a much longer path up to the cathode. Therefore $\Delta t^+ \sim 10$ ms.

If fast pulses are required, one can restrict the response of the detector to the fast electron pulse by introducing an RC-coupling in front of the preamplifier. Choosing $RC \ll \Delta t^+$, the ion pulse is neglected.

If the time constant is chosen to be extremely small, say $RC \approx 1$ ns, then the fine structure of the anode pulse can be resolved. It turns out that the signal consists of several short pulses. Each of these is due to an avalanche and each avalanche originates from a cluster of electrons from the primary ionization process. These clusters slowly drift towards the anode and reach the region of high electric field near the wire one after the other (see fig. 3.3).

The spatial expansion of the electron avalanche leads to a drop-shaped distribution of positive and negative charges near the anode wire. It is caused by the ions lagging behind the fast electrons and moving slowly away from the wire. The shape is depicted in fig. 2.7.

2.3 Geiger–Mueller counters

The spatial localization of the electron avalanche is no longer maintained if, at an electric field strength higher than the one used for the proportional mode, the number of ultraviolet photons formed in the avalanche process increases considerably. If the coefficient describing the frequency of obtaining, in an avalanche, A electrons and γA photoelectrons increases such that $\gamma A \sim 1$, the end of the region of limited proportionality is reached. The ultraviolet quanta propagate in all directions, transverse or parallel to the electric field. They create photoelectrons in the whole gas

Fig. 2.7. Spatial distribution of charges forming an avalanche in a proportional counter; cloud chamber picture (left) and distribution of ions ($+$) and electrons ($-$) (after [LO 61]).

 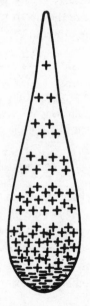

volume and in the walls of the counter. The avalanche spreads over the whole counter and leads to a complete discharge. The charge Q liberated in this process is independent of the primary ionization; it only depends on the capacitance C of the counter and the voltage U_0 applied to the electrodes: $Q = CU_0$. The electric field at which this Geiger–Mueller operation of the counter sets in can be recognized in fig. 2.8 by the observation that the number of ion pairs liberated becomes equal for particles inducing different primary ionization, namely electrons and α-particles. The gas amplification in the Geiger–Mueller region is about $A \sim 10^8$–10^{10}.

Geiger–Mueller counters
If the gas filling of the counter consists of noble gases or two-atomic gases,

Fig. 2.8. Gas amplification factor as a function of voltage applied in a proportional counter for ionizing α-particles and electrons (after [PR 58]).

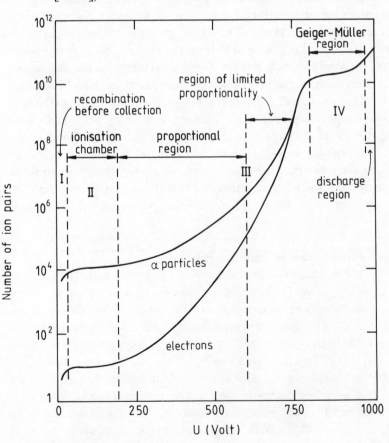

the discharge process is stopped by the cloud of positive ions near the anode wire. These ions reduce the electric field around the wire such that further electrons arriving there cannot form avalanches. As soon as this cloud of ions has migrated away from the anode (after about 1 ms) and has reached the cathode, producing secondary electrons there, the avalanche formation at the anode can start again. The discharge can be quenched externally by choosing the external resistor R sufficiently large, such that the voltage drop due to the anode current I reduces the instantaneous anode potential $U_0 - IR$ below the lower limit required for Geiger–Mueller operation.

Self-quenching counters
The propagation of ultraviolet quanta in the counter can be prevented by adding to the counting gas other gases with large molecules. Most of these quenching agents are organic gases like C_2H_5OH, CH_4, C_2H_6, isobutane (C_3H_8), methylal $(OCH_3)_2CH_2$. These molecules can absorb ultraviolet quanta in the wavelength region $100 \text{ nm} \leqslant \lambda \leqslant 200 \text{ nm}$ and, if they are added in an appropriate concentration, they reduce the range of these photons to the size of the wire diameter. The transverse expansion of the discharge takes place only along the anode wire whereby a tube of positive ions around the wire is formed. Because of their short range, the photons cannot reach the cathode; hence, they cannot produce photoelectrons. In addition, the creation of secondary electrons by positive ions in the cathode is reduced considerably because the ions of the counting gas transfer their ionization to the atoms of the quenching gas, which do not, however, get enough energy for sizeable secondary ionization.

In this way, the discharge stops by itself, and the working resistor R can be chosen much smaller than in the case of the Geiger–Mueller counter. The time constant RC in front of the amplifier can be chosen to be as small as 10^{-6} s.

2.4   Measurement of ionization in liquids

Liquids have some advantages over gases when used as detectors for charged particles. Their density exceeds that of gases by about a factor of one thousand, as does the energy absorbed per unit of the thickness of the layer and the number of primary electrons liberated in the ionization process. The energy needed for creation of an electron–ion pair amounts to $W_i(\text{L Ar}) = 24$ eV in liquid argon and $W_i(\text{L Xe}) \approx 16$ eV in liquid xenon. The statistical fluctuation in the number of ionization electrons is very small: for 1 MeV of absorbed energy in such a detector one expects $N \geqslant 4 \times 10^4$ electrons, and the relative energy resolution from this source should be $\sigma(E)/E = 1/\sqrt{N} < 10^{-2}$. With these materials it is possible to build voluminous and massive detectors at a comparably low cost. The liquids are homogenous, and the charge collection should be reproducible.

In spite of this, detectors with liquid argon have only recently been used successfully in experiments [WI 74, EN 74, KN 74] because electron capture by impurities (mainly oxygen) could only then be controlled by using elaborate purification systems. The mean free path λ_t for electrons before they are captured by impurities ('trapping') is inversely proportional to the concentration k of impurities (e.g. $k = N_{O_2}/N_{Ar}$): $\lambda_t = \alpha E/k$, where E is the electric field and α the trapping constant. For oxygen as an impurity in liquid argon, the constant was measured at fields of $E = 0.2$ to 7.0 MV/m to be $\alpha = (15 \pm 3) \times 10^{-15}$ m^2/V [HO 76]. A reduction of the impurities to a level of $k = 0.2$ to 8.0 ppm, as achieved in experiments, increases the mean free path λ_t to a few millimetres, such that ionization chambers with a distance between electrodes in the millimetre range become possible. Measurements of electron drift velocities in liquid argon and mixtures of liquid argon and methane are shown in fig. 2.9. The electron mobility in

Fig. 2.9. Drift velocity of electrons in liquid argon and argon–methane mixtures [SH 75].

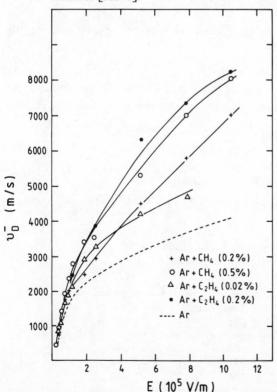

purified liquid argon at a field strength of $E = 1$ MV/m is $\mu_e = 4 \times 10^{-3}$ m^2/V s. The drift velocity of electrons at these high fields is therefore 4×10^3 m/s, about the same as the one in argon gas at $E = 100$ kV/m. At low field strength (10^4 V/m) the electron mobility in liquids is appreciably higher than at high field, as shown in table 9.

On the other hand, the mobility of ions in liquids is very small. At electric fields of $E = 2.4$ to 18.7 MV/m in liquid argon, it amounts to only $\mu_I = 2.8 \times 10^{-7}$ m^2/V s [WI 57]. The pulse induced by the movement of the ions increases, therefore, so slowly that it is not usable for the detection of individual particles. We consider now only the electron component in an ionization chamber by differentiating the pulse with an RC-coupling such that RC is small compared with the ion collection time t_I: $RC \ll t_I$.

We obtain again the pulse shape induced by the moving electrons, but this time we include the case where the path of the particle is *perpendicular* to the electrode of the parallel-plate ionization chamber. As in section 2.1, energy conservation is used. If $q(t)$ is the charge present between the plates and v_D^- the drift velocity of the electrons, we obtain

$$d\left(\frac{Q^2}{2C}\right) = q(t)E_z \, dz = q(t)E_z v_D^- \, dt \tag{2.17}$$

For a track passing at $t = 0$ parallel to the plates at the position $z = z_0$, the charge density is (case I):

$$\rho(z, t) = -Ne \, \delta(z - (z_0 - v_D^- t)) \tag{2.18}$$

and we have the situation of section 2.1 (case I). In liquid argon calorimeters, however, ionizing tracks pass orthogonal to the plates. Then we have (case II):

$$\rho(z, t) = \begin{cases} -\dfrac{Ne}{d} & \text{for } 0 < z < d - v_D^- t \\ 0 & \text{for } \quad z > d - v_D^- t \end{cases} \tag{2.19}$$

The charge between the plates at time t in these two cases is:

Table 9. *Electron mobilities in liquid noble gases at electric field strengths of 10^4 V/m* [BR 79]

Liquid	Temperature (K)	μ_e (m^2/V s)
L Ar	85	0.047
L Kr	117	0.18
L Xe	163	0.22

Case I:

$$q(t) = -Ne \int \delta(z - (z_0 - v_D^- t)) \, dz$$

$$= \begin{cases} -Ne & \text{for } t < t_0 = \dfrac{z_0}{v_D^-} \\ 0 & \text{for } t > t_0 \end{cases}$$

Case II:

$$q(t) = \begin{cases} -\dfrac{Ne}{d}(d - v_D^- t) & \text{for } t < t_d = \dfrac{d}{v_D^-} \\ 0 & \text{for } t > t_d \end{cases}$$

Energy conservation from eq. (2.17) leads to:

$$\frac{2Q_0 \, dQ}{2C} = E_z v_D^-(-Ne) \, dt \qquad \text{(case I)} \quad \text{for } t < t_0$$

$$= E_z v_D^- \left(-\frac{Ne}{d}(d - v_D^- t) \right) dt \quad \text{(case II)} \quad \text{for } t < t_d$$

and after integration one obtains:

Case I:

$$dQ = -\frac{CE_z}{Q_0} v_D^- Ne \, dt$$

$$= -\frac{Ne \, dt}{t_d}$$

$$Q(t) - Q_0 = \begin{cases} -\dfrac{Ne\,t}{t_d} & \text{for } 0 < t < t_0 \\ -\dfrac{Ne\,z_0}{d} & \text{for } \quad t > t_0 \end{cases} \tag{2.20}$$

Case II:

$$dQ = -\frac{CE_z}{Q_0} v_D^- Ne \left(1 - \frac{v_D^-}{d} t \right) dt$$

$$Q(t) - Q_0 = \begin{cases} -Ne\left(\dfrac{t}{t_d} - \dfrac{1}{2}\left(\dfrac{t}{t_d}\right)^2\right) & \text{for } 0 < t < t_d \\ -\dfrac{Ne}{2} & \text{for } \quad t > t_d \end{cases} \tag{2.21}$$

For liquid argon chambers with a gap d between plates we therefore get from case II by differentiation:

$$i(t) = -Ne \frac{1}{t_d}\left(1 - \frac{t}{t_d}\right) \quad \text{for } t < t_d \tag{2.22}$$

The time interval $t_d = d/v_D^-$ for $d = 2$ mm and $v_D^- = 4 \times 10^3$ m/s is approximately 0.5 μs. These liquid argon chambers are used mainly in total absorbing calorimeters for electromagnetic or hadronic showers (see sections 6.1 and 6.2).

2.5 Semiconductor detectors

Semiconductor detectors work as solid-state ionization chambers. A charged particle or, in the case of γ-ray detection, a photoelectron creates electron–hole pairs on its way through a crystal. The crystal is situated between two electrodes generating an electric field. During the ionization process, electrons from the valence band are excited by the passing particle and are transferred to the conduction band, leaving behind a hole in the valence band. In the secondary process, the electrons dissipate their remaining energy by producing more electron–hole pairs (excitons) and by exciting lattice oscillation (phonons). After the passage of the charged particle there remains a tube of plasma around the track with high concentrations of electrons and holes (10^{15} to $10^{17}/\text{cm}^3$). The problem is now to separate these charges and collect the electrons on the anode before they recombine with the holes. This number of electrons is larger than in gases for the same amount of energy deposited in the material. For the formation of one electron–hole pair in silicon (germanium) only ionization energies of 3.6 eV (2.9 eV) are needed, compared with values between 20 and 40 eV in gases. In scintillation counters, as much as 400 to 1000 eV are required in order to liberate, via the scintillation light falling onto a photocathode, one photoelectron.

For these solid-state counters, ultrapure semiconductor materials are used. Their resistivity reaches $5 \times 10^4 \,\Omega$ cm for silicon and $50 \,\Omega$ cm for germanium. These semiconductor crystals are operated as diodes with reverse bias in order to generate high electric fields for the collection of charges. For practical purposes, three types of semiconductor detectors have been used: diodes with a p–n junction, surface barrier junctions, and p–i–n junction detectors.

A *p–n junction* in semiconductors is the border plane between a p-type zone in the crystal doped with a uniform concentration of electron–acceptor impurities, and an n-type zone doped with electron–donor impurities [SH 50, BU 60, DE 66, BE 68]. An asymmetric p–n junction consists of weakly doped p-silicon into which a layer of high concentration n-type impurities is introduced by diffusion from one surface. At the boundary between the two types of material a double layer of charges is formed in the following way: conduction electrons from the n-type side migrate into the p-type material, and holes diffuse across the junction into the n-type zone. At the same time, the electrons leave behind on the n-doped side ionized (positive) donor impurities, and the holes moving to the n-doped side leave acceptor impurities with an extra electron on the p-side. As a net effect, there is a surplus of positive space charge on the n-side and of negative space charge on the p-side of the junction. The build up of this double layer is slowed down by the electric field created by the space charges until an equilibrium is reached. At equilibrium, this steady electric field pushes back free electrons into the n-type zone and holes into the p-type zone, such that the layer over which this charge imbalance exists is depleted of free charge carriers. In this 'depletion layer' the density of charge carriers is $10^2/cm^3$ compared with $10^{10}/cm^3$ for the purified silicon.

The potential difference between the two sides of a p–n junction due to the double layer of space charges can also be visualized in the energy band picture: in an n-doped semiconductor the Fermi energy level is higher relative to the upper edge of the valence band than for a p-type material. When merging together the two types of material, the Fermi energy level must be the same on both sides of the boundary; therefore the bands must be shifted in the way shown in fig. 2.10. If an external potential U is applied in the reverse direction of the diode, i.e. a negative potential applied to the p-type side, the potential difference due to the space charges is enhanced. Since the resistivity of the depletion layer is much higher than the one of the remaining parts of the junction material, virtually all the reverse bias voltage is present across the depletion layer, thus increasing greatly the electric field there. One can use Poisson's equation between the potential $U(x)$ and the charge density $\rho(x)$ in one dimension:

$$\frac{d^2 U(x)}{dx^2} = \frac{-\rho(x)}{\varepsilon_0 \varepsilon} \tag{2.23}$$

and with $E_x = -dU/dx$

$$\frac{dE_x(x)}{dx} = \frac{\rho(x)}{\varepsilon_0 \varepsilon} \tag{2.24}$$

in order to see that the space charges also increase by the application of the reverse bias, and that the thickness of the depletion layer increases as well. Let N_D and N_A be the density of donor and acceptor impurities, respectively, and the asymmetric double layer of charges be represented by the simplified form

$$\rho(x) = \begin{cases} eN_D & \text{for } -a < x \leqslant 0 \\ -eN_A & \text{for } 0 < x \leqslant b \end{cases} \tag{2.25}$$

with $N_D \gg N_A$ and $a < b$. For solving the Poisson equation we require the boundary conditions for the electric field to vanish outside the space charge layer:

$$E_x(-a) = -\frac{dU}{dx}\bigg|_{-a} = 0 = E_x(b) = -\frac{dU}{dx}\bigg|_{b} \tag{2.26}$$

The first integration of eq. (2.23) then gives

$$\frac{dU}{dx} = \begin{cases} -\dfrac{eN_D}{\varepsilon_0\varepsilon}(x+a) & \text{for } -a < x \leqslant 0 \\[2ex] +\dfrac{eN_A}{\varepsilon_0\varepsilon}(x-b) & \text{for } \quad 0 < x \leqslant b \end{cases} \tag{2.27}$$

For the potential itself, we have the boundary conditions (neglecting the potential due to the original junction space charge)

$$U(-a) = 0 \quad \text{and} \quad U(b) = -U_0$$

Fig. 2.10. Band structure in an asymmetric p–n junction. d: thickness of depletion layer; E_c: lower limit of conduction band; E_v: upper limit of valence band; U: reverse bias voltage. (After [BR 61].)

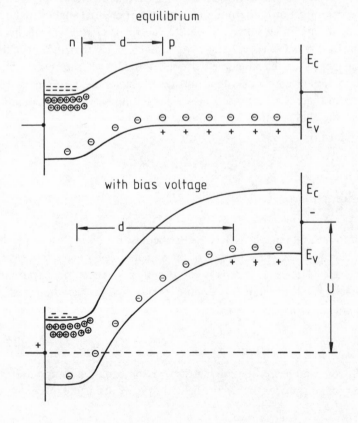

The second integration then yields

$$U(x)= \begin{cases} -\dfrac{eN_D}{\varepsilon_0\varepsilon}(x+a)^2 & \text{for} \quad -a<x\leqslant 0 \\[2ex] \dfrac{eN_A}{\varepsilon_0\varepsilon}(x-b)^2 - U_0 & \text{for} \quad 0<x\leqslant b \end{cases} \tag{2.28}$$

The solution for $U(x)$ must be continuous at $x=0$, and the charges in the double layer compensate, i.e. $N_D a = N_A b$, which together with eq. (2.28) yields

$$b(a+b)=\frac{2\varepsilon_0\varepsilon U_0}{eN_A} \tag{2.29}$$

Because $N_D \gg N_A$, the thickness of the depletion layer b on the p-type side is much larger than a, and therefore the total depth of the depletion layer is $d=a+b\approx b$. Therefore

$$d\approx \sqrt{\left(\frac{2\varepsilon_0\varepsilon U_0}{eN_A}\right)} \tag{2.30}$$

Noting that the concentration of impurities is inversely proportional to the resistivity r of the basic material (p silicon in our case) multiplied by the mobility μ of the charge carriers, $1/eN_A = r_p\mu$, another form of this relation is

$$d\approx \sqrt{(2\varepsilon_0\varepsilon U_0 r_p \mu)} \tag{2.31}$$

The highest electric field occurs at $x=0$; according to eqs (2.27) and (2.30):

$$E_x(0)= \sqrt{\left(\frac{2eN_A U_0}{\varepsilon_0\varepsilon}\right)} = \frac{2U_0}{d} \tag{2.32}$$

This field, for $d=100\ \mu$m and $U_0=200$ V, is 4×10^6 V/m and is sufficient to separate a large part of the electrons and holes produced by the incident radiation and collect them on the electrodes before they can recombine. The collection times t_c can be estimated; for a counter thickness of $d=1$ mm, an *average* electric field of $E=2\times 10^5$ V/m and an electron mobility $\mu=2\times 10^4$ cm^2/Vs it is given by

$$t_c=\frac{d}{\mu E}=2 \text{ ns} \tag{2.33}$$

Similarly for the p–n junction detector, a depletion layer can also be produced by starting with an n–type silicon crystal and generating a p-type layer by etching the surface and then evaporating on it a thin gold layer under oxidizing conditions. This type of counter is called a *surface barrier counter* and behaves in much the same way as a diffused p–n junction. Counters with areas up to 10 cm^2 and depletion layer thickness of 50 μm or with areas of 1 cm^2 and thickness up to 2 mm can be manufactured.

For both types of counters the thickness d of the depletion layer increases with the reverse bias voltage U_0 in the way given in eq. (2.31). Putting in numbers, one gets

$$d = \begin{cases} 0.309 \sqrt{(U_0 r_\mathrm{p})} & \text{for p-type Si} \\ 0.505 \sqrt{(U_0 r_\mathrm{n})} & \text{for n-type Si} \end{cases} \tag{2.34}$$

where d is measured in micrometres, U_0 in volts, and the resistivity r in ohm centimetres. These equations can be represented graphically in a diagram as shown in fig. 2.11 [BL 60].

Fig. 2.11. Diagram representing the relation between the thickness of the depletion layer in a silicon junction, the bias voltage and the resistivity [BL 60].

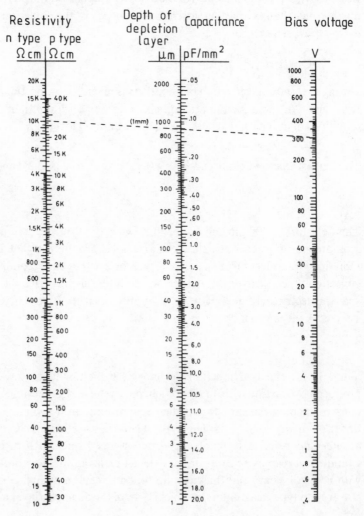

The thickness of depletion layers is limited for these junction types to less than 2 mm. If thicker sensitive layers are required, lithium-drifted p–i–n detectors can be used. In these detectors, a very thick depletion layer is achieved by producing in the semiconductor crystal between an n-type and a p-type zone a layer in which the impurities have been almost completely compensated by ions of the other electron affinity. These ions are introduced into the crystal by a special process of ion drifting. In particular, one usually starts with the semiconductor material of highest available purity, which is p-type and has remaining impurities of boron. From one surface of the crystal lithium ions (donor impurities) are diffused into the crystal, thus forming an n-type layer. A reverse bias is then applied to the junction, and the temperature raised to about 70 °C. The lithium ions slowly migrate into the p-type region and are able to compensate completely the remaining p-impurities there. The process takes from days to weeks until exact compensation is achieved. The mobility of lithium ions is very large in germanium, where depletion layer thicknesses up to 20 mm can be achieved. However, the mobility is so high at room temperature that a Ge(Li) counter has to be cooled to liquid nitrogen temperature (77 K) in order to preserve the lithium distribution and therefore the depletion layer. For Si(Li) counters this is not necessary because of the lower lithium ion mobility in silicon.

In the region in which the lithium ions compensate the remaining p-impurities, the resistivity of the material increases to the *intrinsic* value of the semiconductor without any impurities, i.e. $3 \times 10^5 \ \Omega$ cm for silicon and $50 \ \Omega$ cm for germanium. The density of uncompensated impurities can be decreased to about $10^9/\text{cm}^3$. This region is therefore called the layer of 'intrinsic conductivity', and the total detector is called a p–i–n junction.

If we consider the range diagram of fig. 2.12, we see that in a sensitive layer of 5 mm of silicon, α-particles with energies up to 200 MeV and electrons with energies up to 2 MeV can be absorbed. In this energy range the charge collected in the junction is approximately proportional to the particle energy E_0.

In this range, the energy resolution of semiconductor counters is better than that of other detectors. The number of liberated electron–hole pairs is $n = E_0/W_i$ with $W_i = 3.6$ eV (2.8 eV) for Si (Ge). The statistical fluctuation of this number, $\sqrt{n}$, is reduced by the Fano factor F, such that $\sigma_n = \sqrt{(nF)}$. Values for this factor at 77 K are 0.09 to 0.14 for silicon and 0.06 to 0.12 for germanium. The relative energy resolution from statistical fluctuations alone is then

$$\frac{\sigma(E)}{E_0} = \sqrt{\left(\frac{F W_i}{E_0}\right)} \qquad (2.35)$$

Fig. 2.12. The range R of different charged particles in silicon as a function of their energy E (after [RI 54]).

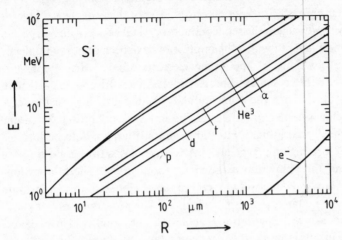

Therefore, for a germanium counter one expects, as a lower limit to the relative resolution for a photon of 8 MeV, $\sigma(E)/E_0 \gtrsim 2 \times 10^{-4}$, and for a photon of 122 keV, $\sigma(E)/E_0 \geq 1.7 \times 10^{-3}$. In practice, resolutions of 5.4×10^{-4} and 7.1×10^{-3} have been measured, which shows that other factors deteriorate the resolution, like incomplete charge collection and electronic noise.

3

....

Measurement of position

3.1 Multiwire proportional chamber

In a proportional chamber, the principle of a proportional counter (section 2.2) is applied to a detector with large area. The mechanical arrangement and the electric field configuration are shown in fig. 3.1. A series of parallel anode wires is stretched in the middle plane between two cathode planes made of metal foil, a wire mesh or wires. A charged particle traversing the three planes leaves behind a string of electrons and ions formed by the ionization process. Electrons travel along the electric field lines towards one of the anode wires. The field strength *near* the wire is approximately the same as in a cylindrical capacitor, eq. (2.4), as shown in fig. 3.1. When the kinetic energy gained by the electron between two collisions (eq. (2.9)) exceeds the ionization threshold of one of the components of the counting gas, secondary ionization and avalanche formation sets in, as in the proportional counter, and a pulse is induced on the anode wire L near which the avalanche formation has taken place. The main contribution to the pulse seen on the anode wire L is induced by the ions in the avalanche moving away from the wire. This pulse consists of several individual pulses each due to a separate avalanche. These avalanches are triggered by clusters of electrons from the primary ionization process which drift into the region of high field strength one after another.

The time sequence of avalanche formation on the anode wire L is shown in fig. 3.2. While the electron cloud formed in the avalanche develops into a cylindrical tube around the anode wire, the shape of the slowly moving ions still retains the memory of the origin of the avalanche by forming a drop-like shape. The time structure of pulses on the anode wire due to one primary ionization process is revealed in fig. 3.3 with an oscilloscope trace of good time resolution [FI 75]. The individual pulses from separate

Fig. 3.1. Principle of multiwire proportional chamber. Upper part: schematic of geometry; lower part: equipotential surfaces (dashed) and electric field lines (full curves) in the neighbourhood of two anode wires in the plane perpendicular to the wire direction [ER 72].

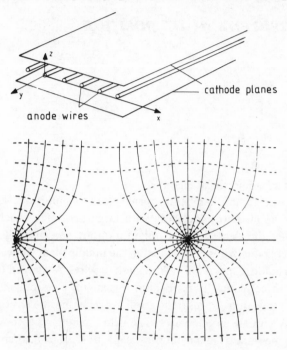

Fig. 3.2. Time development of an avalanche near to an anode wire in a proportional chamber. (a) Primary electron moving towards anode. (b) The electron gains kinetic energy in the electric field and ionizes further atoms; multiplication starts. (c) The electron and ion clouds drift apart. (d), (e) The electron cloud drifts towards the wire and surrounds it; the ion cloud withdraws radially from the wire [CH 72].

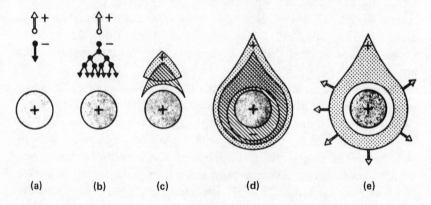

(a)   (b)   (c)   (d)   (e)

avalanches have a risetime of about 0.1 ns and a decay time determined by the time constant RC of the differentiating circuit.

According to Charpak *et al.* [CH 68], the anode wires of the multiwire proportional chamber act as independent detectors. The capacitive coupling of the negative pulses on wire L to a neighbouring anode wire A is compensated by the positive pulse induced on A by the ions in the avalanche moving away from L and towards A.

For the mechanical construction of proportional chambers, one rule is to

Fig. 3.3. (a) Oscilloscope picture of voltage pulse from a proportional chamber. (b) Computer simulation of such a pulse [FI 75].

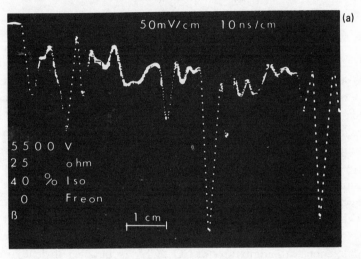

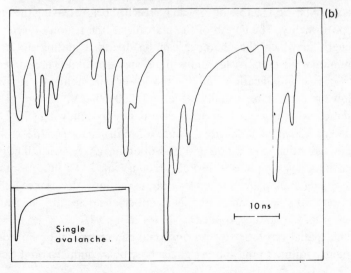

use a wire diameter of about 1% of the distance between anode wires to achieve sufficiently large electric fields for gas amplification, e.g. 20 μm diameter for 2 mm distance. The standard material is gold-plated tungsten wire for the anodes. The frames with printed circuit boards are usually made of glass fibre material. Cathode planes are stretched metal foils or are made of wires (Cu–Be, dia. 50–100 μm).

A problem specific to large chambers is the mechanical instability of anode wires due to their electrostatic repulsion. The computation of this effect [TR 69] shows that the wires are stable if the wire tension T surpasses a limiting value T_0 given by the chamber geometry and the potential difference V between anode and cathode:

$$T > T_0 = \left(\frac{Vl}{2\pi a}\right)^2 4\pi\varepsilon_0 \tag{3.1}$$

Here l is the length of the anode wire and a the half-gap between anode and cathode. Putting in numbers, one obtains for $V = 4.3$ kV, $a = 6$ mm and $l = 60$ cm a minimum tension of $T_0 = 0.5$ N. This approximates to the tension applicable to 20 μm tungsten wires (elastic limit around 0.8 N). For chambers exceeding these dimensions, the anode wires have to be supported every 60 cm. Several solutions to this electromechanical problem have been tested. One possibility is to weave a thin nylon wire across the anode wires, perpendicular to their direction. The disturbance of the electric field near the anode wires then causes an inefficient zone of 5 mm width along the nylon wire [KL 70].

Counting gases for proportional chambers include the noble gases argon and xenon with admixtures of CO_2, CH_4, isobutane, ethylene or ethane. Gas amplification factors of 10^5 can be reached before leaving the region of proportionality. The length of the plateau, i.e. the region of electric field strength in which the chamber can be operated at full efficiency for minimum ionizing particles and without spark breakdown, is determined by the gas amplification and by the lower threshold of the amplifier following each wire, usually at about 200–500 μV. A typical set of parameters for large (> 1 m^2 area) chambers is the following [SC 71]: wire diameter $= 20$ μm; anode wire distance $= 2$ mm; distance anode–cathode $= 6$ mm; counting gas argon (80%)–isobutane (20%) with methylal as quenching gas; amplifier threshold $= 200$ μV on 2 kΩ; time resolution $= 30$ ns; length of plateau $= 700$ V; spatial resolution (variance of a box distribution) $\sigma_x = 0.7$ mm. For this chamber, the detection efficiency as a function of the voltage applied is shown in fig. 3.4.

The spatial resolution of proportional chambers can be improved by using in addition to the anode signals those pulses induced on the cathode plane [RA 74, CH 78*a*]. For this purpose, one cathode plane is made of metal strips or groups of connected wires running in orthogonal direction to the anode wires (fig. 3.5). On these strips, pulses are induced by the avalanches occurring near an anode wire. The magnitude of the pulses induced on the individual strips varies with the distance between the avalanche and the strip; the centre of gravity of the charges induced on the strips can be determined: it is a precise measurement of the avalanche position. A position resolution of $\sigma_x \sim 35$ m in the direction transverse to the strip direction can be achieved. Fig. 3.6 shows such a measurement

Fig. 3.4. Detection efficiency of a proportional chamber, with 6 mm gap as a function of the high voltage, for different lengths of the electronic gate for accepting pulses [SC 71].

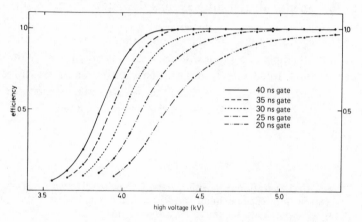

Fig. 3.5. Principle of cathode readout in proportional chambers. The centre of gravity of charges induced on the cathode strips running orthogonal to the anode wires determines the position of the avalanche (a) [CH 78*a*].

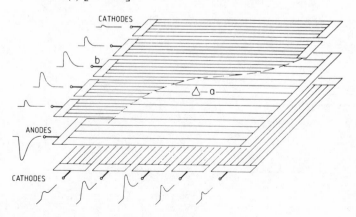

using soft X-rays as a radiation source. The source is located in three positions, each 200 μm apart from another. It produces, via photoelectrons, localized ionization. The centre of gravity of the induced charges is well separated for the three positions, and the experimental resolution $\sigma_y =$ 35 μm is not even the lower limit for this cathode readout method because the finite range of the photoelectrons also contributes to σ_y.

Proportional chambers with cathode readout thus yield an excellent spatial resolution. However, the expenditure for mechanical construction and electronic pulse processing is considerable.

3.2 Planar drift chambers

The drift chamber [WA 71, CH 70*a*] is based on the observation that there is an unequivocal correlation of the time difference Δt between the passage of the particle and the rising edge of the anode pulse with the distance between the point of primary ionization and the anode wire in a proportional chamber. This time difference Δt is mainly given by the drift time of electrons formed at $t = t_0$ in the primary ionization process up to the moment t_1 where they enter the high field region around the anode wire and generate avalanches. The drift path of the electrons is then

$$z = \int_{t_0}^{t_1} v_D(t)\, dt \tag{3.2}$$

Fig. 3.6. Illustration of position resolution achievable with cathode strip readout. The chamber was irradiated with a 1.4 keV X-ray source at three positions, separated by 200 μm; y is the centre of gravity of charges on the strips [CH 78*a*].

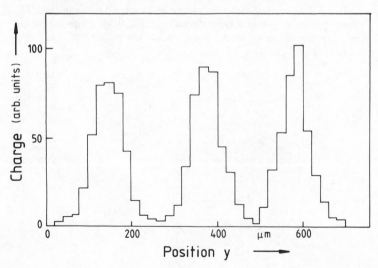

It is desirable to have a constant drift velocity v_D along the drift path because then eq. (3.2) becomes a linear relation:

$$z = v_D(t_1 - t_0) = v_D \, \Delta t \qquad (3.3)$$

For a constant drift velocity of 50 mm/μs a time measurement with a precision of 4 ns can give a spatial resolution of $\delta z = 200$ μm. Constancy of v_D can be achieved by keeping the electric field strength along the drift path constant. In a conventional proportional chamber with parallel anode wires between the two cathode planes this is not possible because of the zero field region between the two anode wires. By introducing a field wire at a potential $-HV1$ between two anode wires at $+HV2$, Walenta, Heintze and Schuerlein [WA 71] succeeded in obtaining an almost linear relation between drift time and drift path.

A complete linearization of this relation can be attained by making the electric drift field as constant as possible over the whole drift space. Different electrode configurations serve this purpose. The drift cell of a planar chamber shown in fig. 3.7 has one anode wire at $+HV2$ and two field wires at $-HV1$ closing up the cell. The field strength in the drift region far from the anode wire is kept constant by varying the potential on the cathode wires linearly from zero at the wire opposing the anode wire to $-HV1$ at the wire opposite to the field wire.

Another possibility of arriving at constant fields involves using cathode planes at a fixed potential but shaping the potential inside the cell by an appropriate number of field wires. In the cell structure shown in fig. 3.8, there are six field wires and ten corrector wires per anode wire, in a cell with a cross-section of 60×30 mm^2. Fig. 3.9 shows the equipotential lines for

Fig. 3.7. Schematic of a drift chamber cell. Equipotential lines are drawn as full lines; $+HV2$: potential of anode wire; $-HV1$: potential of cathode wires; other field wires: potential varying between 0 and $-HV1$.

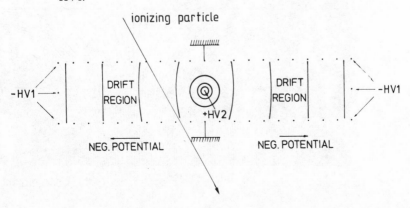

this cell, and fig. 3.10 demonstrates a nearly linear relation between drift path and drift time, as measured with minimum ionizing particles.

Apart from the field distribution, the dependence of the drift velocity v_D on the electric field strength E also has an influence on the linearity of the drift path–drift time relation. For some specific mixtures (e.g. argon–

Fig. 3.8. Cell structure of large planar drift chamber. S.W.: signal wire (anode); other wires are potential wires [MA 77].

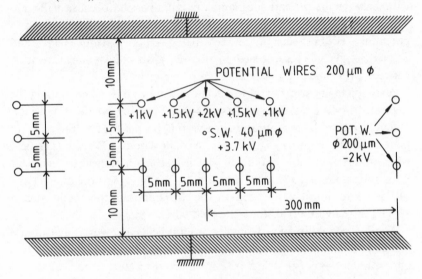

Fig. 3.9. Equipotential lines in one quadrant of the cell shown in fig. 3.8 [MA 77]. Curves are equipotential lines; arrows indicate drift direction of electrons.

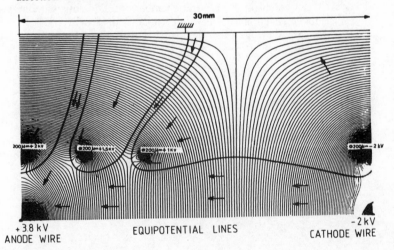

isobutane 80% : 20%) v_D is nearly independent of E in the domain $E = 0.5$–2 kV/cm (see fig. 1.8), such that even remaining inhomogeneities in the electric field only weakly influence the linearity of the relation.

Large planar drift chambers consist of many (~ 100) such drift cells. The spatial resolution attainable with large area ($\gtrsim 10 \text{ m}^2$) chambers is mostly limited by the mechanical tolerances in positioning the anode wires (100–300 μm) and the sagging of wires under their own weight. For smaller chambers, however, the limits of spatial resolution are determined by the time resolution of the time-to-digital converters ($\geqslant 40 \ \mu$m) and the diffusion of electrons during their migration towards the anode ($\leqslant 50 \ \mu$m for a drift path of 20 mm).

3.3 Cylindrical drift chambers

For experiments performed in colliding beam machines, detectors with solenoidal magnetic fields around the interaction point are often used.

Fig. 3.10. Relation between drift time t and drift distance for cell shown in fig. 3.8 [MA 77].

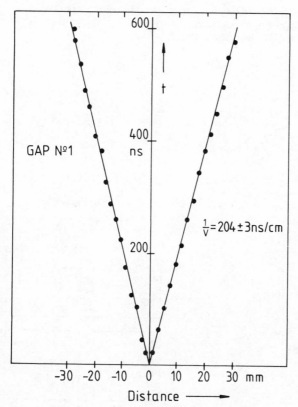

In cylindrical co-ordinates the field components are $B_r = B_\varphi = 0$ and $B_z \neq 0$. The cylindrical symmetry of the field then also requires position detectors with such a geometry. These detectors measure the initial direction of particles emerging from the interaction point and their radius of curvature in the magnetic field. Geometrical configurations of four such cylindrical detectors are sketched in fig. 3.11.

The first detectors of this kind made use of cylindrical layers of proportional chambers (section 3.1) or spark chambers (section 3.9) to measure the track positions (fig. 3.11a). The anode wires of the proportional chambers are stretched parallel to the magnetic field **B**; the electric field **E** is radially oriented. The electrons drifting in the electric field are deflected by the Lorentz force in the φ-direction which introduces an error which is, however, small in this case because the drift paths are short, usually below 10 mm. For the second generation of such track detectors, cylindrical drift chambers were used. They consist of up to 20 cylindrically symmetric layers of drift cells. The electric field lines lie in the (r, φ)-plane perpendicular to the

Fig. 3.11. Four different types of cylindrical wire chambers: (a) Proportional chambers; (b) cylindrical drift chambers; (c) Jet-drift chamber; (d) time projection chamber.

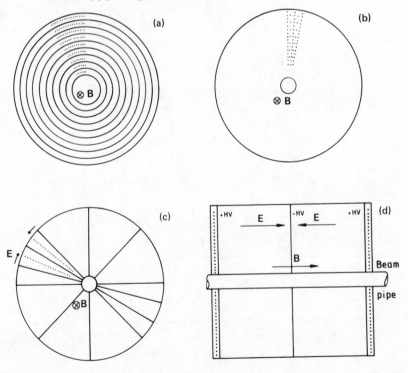

axial magnetic field (fig. 3.11*b*). This field is generated by a suitable arrangement of potential wires which are parallel to each other and surround the signal (anode) wire in the centre of the cell. About half of the signal wires are parallel to the **B**-field, while the others are skew and run at a stereo angle γ (e.g. $\gamma = \pm 4°$) relative to this axis. This enables the reconstruction of the *z*-position of the track, although with limited precision. In order to minimize the number of potential wires, one can omit the closure of the drift cell in radial direction ('open cell geometry'). Thus the homogeneity of the electric field and the linearity of the drift path–drift time relation deteriorate.

An example of such a cylindrical drift chamber with open cell geometry is the TASSO chamber [BO 80*a*]. Fig. 3.12 shows a transverse cut through some drift cells, the length of the wires is 3.5 m. Fifteen cylindrical layers of drift cells allow the measurement of tracks over a radial length of 85 cm; in six of these layers the signal wires form a stereo angle of $\gamma = \pm 4°$ with the axis. The space resolution in the (*r*, φ)-plane is $\sigma_r \sim 200$ μm, the one in the *z*-direction is approximately $\sigma_{r\varphi}/\sin \gamma \sim 3$ mm. In fig. 3.13 the (*r*, φ)-projection of tracks from an event in which an electron and a positron of 15 GeV energy each have annihilated into at least 12 charged particles.

Another example of a cylindrical drift chamber is the ARGUS chamber built for an experiment at the e^+e^- storage ring DORIS II [DA 83]. Here the drift cell is closed (fig. 3.14), and the arrangement of the ten potential wires of one cell around the signal wire produces a nearly symmetric electric field, such that surfaces of equal drift time have a cylindrical shape around the signal wire. This property facilitates pattern recognition, i.e. the

Fig. 3.12. Geometrical arrangement of axial wires in the TASSO cylindrical drift chamber; dimensions are in millimetres [BO 80*a*].

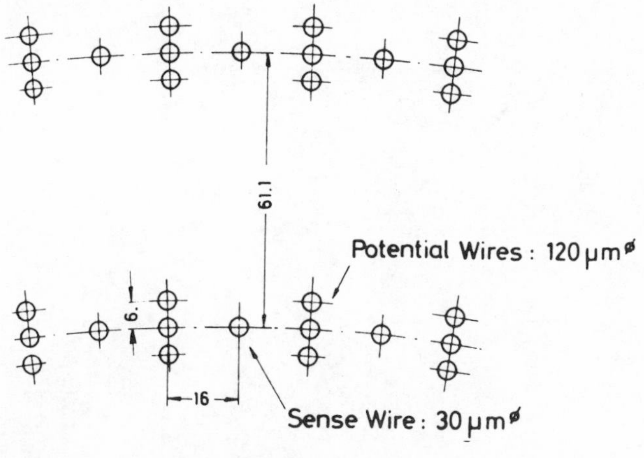

problem of finding the 'best' continuous tracks through the positions of particle traversal measured in the individual drift cell layers at different radial distances from the axis. The spatial resolution achieved with this chamber is $\sigma_{r\varphi} = 190 \ \mu m$ [DA 83].

The two other cylindrical chambers shown in fig. 3.11, the Jet drift chamber (fig. 3.11c) and the time projection chamber (fig. 3.11d) are treated in the following chapters.

3.4   The Jet drift chamber

With this type of cylindrical drift chamber, the number of measured points along a radial track is considerably increased compared with the two chambers discussed in section 3.3. About one measurement per centimetre of track length is recorded. This is achieved by the geometrical

Fig. 3.13. Axial projection of particle track co-ordinates in the TASSO drift chamber; the particles originate from an e^+e^- reaction at the beam crossing point ($+$) at a centre-of-mass energy of 30 GeV [BO 80a].

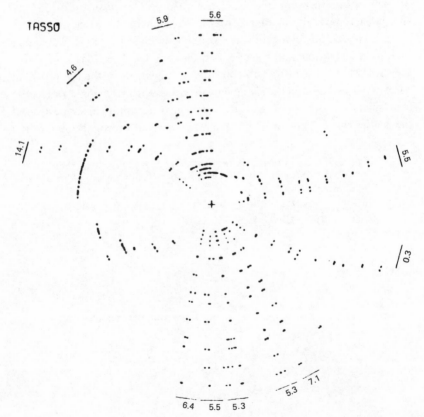

structure shown in fig. 3.11c: the signal wires run in an axial direction, forming the middle plane of a sector-shaped cell with the transverse electric field in the φ-direction. The drift paths of electrons in the gas are much longer than in the cylindrical drift chamber of fig. 3.12 (they can be as long as 10 cm).

The first such Jet chamber was constructed for the JADE detector at the electron–positron storage ring PETRA [BA 79, DR 80]. The cylindrical volume of the chamber is subdivided into 24 radial segments with an opening angle of 15°. In each segment (fig. 3.15) are contained 64 signal wires parallel to the magnetic field, arranged in four cells with 16 wires each. They measure a radial track over 57 cm length. In total, the chamber has 1536 signal wires of 234 cm length. The electric field is transverse to the signal wires, i.e. to the magnetic field direction.

Because the drift paths of electrons are long in this chamber compared with those in traditional drift chambers, the influence of the Lorentz force $e\mathbf{v} \times \mathbf{B}$ becomes noticeable here. As discussed in section 1.3.5, the drift

Fig. 3.14. Geometrical layout of wires, drift paths of electrons towards the central anode wire (full lines) and lines of equal drift time (dashed lines) in a cell of the ARGUS drift chamber; magnetic field of 0.9 T parallel to the wires [DA 83].

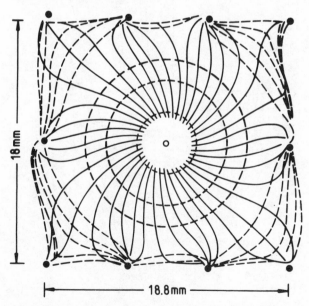

18 mm

18.8 mm

o Signal Wire
• Potential Wire

direction of electrons deviates from the direction of the electric field **E**, by an angle α_L given in eq. (1.73), the Lorentz angle. It depends on the magnetic field B, the electric field E and the drift velocity v_D [WA 81b]:

$$\tan \alpha_L = \omega\tau \approx k(E)v_D \frac{B}{E} \tag{3.4}$$

and a typical value for the JADE chamber with $B = 0.45$ T and a gas pressure of 4 bar is $\alpha_L = 18.5°$. The factor $k(E)$ depends on the gas mixture and the electric field E [SC 80]. Because of this effect, the surfaces of equal drift time to the signal wire are complicated in the neighbourhood of this signal wire (fig. 3.16). In the close vicinity of the wire (distance less than 5 mm) they are cylindrical surfaces; at larger distances the figure shows the shape of the distorted surfaces.

Fig. 3.15. Cross-section through two segments of the Jet chamber in the JADE experiment. The radial length of the three layers is 57 cm. Length of drift path d; Lorentz angle α [DR 80, WA 81b].

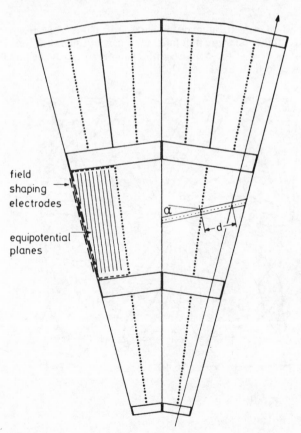

Fig. 3.17 shows the (r, φ)-projection of a two-jet event produced in an electron–positron annihilation at an energy of 35 GeV in the centre-of-mass system. Up to 48 space co-ordinates are measured along a radial track. The signals can also be used for a measurement of ionization energy loss of the track if the pulse height information on the wire is recorded. The z-co-ordinates of the track elements are determined by measuring the charge induced on both ends of the signal wire. From the ratio of these charges, the z-co-ordinate of the track segment can be obtained with an uncertainty of ± 1.6 cm.

Chambers of this type are sometimes called 'pictorial drift chambers'. They have been used for the central detectors of the axial field spectrometer (AFS) [CO 81] and of the UA1 detector [BA 80]. A collection of properties of such central detectors can be found in table 20.

3.5 Time projection chamber (TPC)

A novel way of using the proportional and drift chamber principles for the central detector of a storage ring was proposed by D. R. Nygren [NY 74, NY 81]. Fig. 3.18 shows a large cylindrical volume with radius 1 m and length 2 m filled with an argon–methane mixture at a gas pressure of

Fig. 3.16. Drift path of electrons (full lines) and lines of equal drift time (dashed) near the anode wire in the Jet chamber of JADE [DR 80, WA 81b].

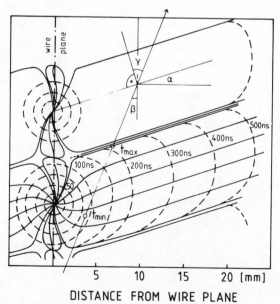

DISTANCE FROM WIRE PLANE

Fig. 3.17. Axial projection of tracks from an e^+e^- interaction at 30 GeV centre-of-mass energy in the Jet chamber of JADE. Hits associated with a track are shown as (solid +), other hits as (faint /) [DR 80].

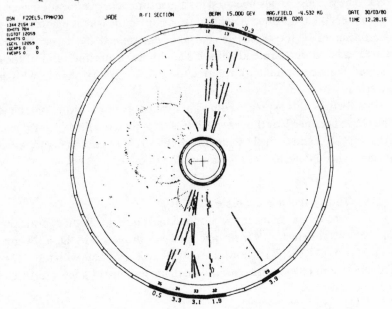

Fig. 3.18. Sketch of time projection chamber (TPC) for the PEP4 experiment [MA 78].

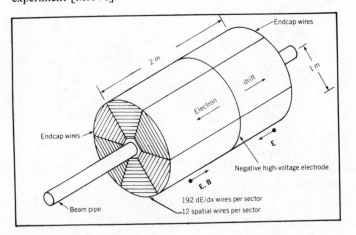

10 bar. The two end caps of the barrel are equipped with one layer of multiwire proportional chambers subdivided into six sectors. Each of the sectors has 183 proportional signal wires for multiple measurements of ionization along a radial track. Of those, 15 wires have a segmented cathode readout ('pads', see section 3.1) for the spatial measurement mainly of the azimuthal φ-co-ordinate, while the r-co-ordinate is determined mainly by the wire position (fig. 3.19).

The construction principle is different from that in other chambers by the fact that the electric drift field (150 kV/m) and the magnetic field of the solenoid ($B = 1.5$ T) are parallel. By this arrangement, Lorentz forces on the drifting electrons vanish. The drift direction is parallel to the **E** and **B** fields. The electrons formed by an ionizing particle emerging from the interaction point at the centre of the barrel drift towards one of the end caps. In this process, the image of the ionizing track is broadened by the transverse diffusion of electrons during the drifting. However, the strong magnetic field reduces considerably this widening of the track image by forcing the electrons to perform helical movements around the magnetic field lines (see section 1.3.6). The transverse diffusion coefficient is reduced by a factor of $1/(1+\omega^2\tau^2)$, where $\omega = eB/m$ is the cyclotron frequency of electrons and τ is the mean free time between two collisions.

The spatial reconstruction of the ionizing tracks is performed by measuring the two-dimensional projected images in the (r, φ)-plane on the end caps using the cathode pad readout and the centre-of-gravity method, and by registering in addition the arrival time of drifted electrons at the end cap wires giving the z-co-ordinate. In this way, three-dimensional track

Fig. 3.19. Principle of track measurement by cathode pad readout in a TPC [FA 79].

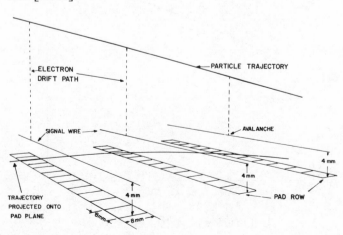

segments are obtained, and pattern recognition and track finding are alleviated considerably compared with chambers with uncorrelated co-ordinate information. The measured drift time gives the z-co-ordinate of the track with a precision of $\sigma_z \sim 200 \, \mu m$ – more accurately than in other cylindrical chambers. The spatial resolution in the (r, φ)-plane is about 180 μm if the chamber is operated at low counting rate, e.g. with cosmic muons. At higher counting rates the resolution deteriorates due to the field distortions from the space charge of ions formed in the avalanches near the counting wires (fig. 3.20). The density of such positive ions is $\sigma_+ = fAN/(ld\mu_+ E_0)$, where f is the fraction of positive ions migrating back to the drift region, A the gas amplification at the anode wires, N the number of ionization electrons reaching a signal wire per unit of time, l the wire length, d the wire distance, μ_+ the mobility and E_0 the drift field strength. This density can be greatly reduced by introducing an additional wire plane between the proportional chamber and the drift volume which prevents the ions from migrating back to the drift region. This 'gate' is opened for a short time to allow the passage of electrons from the drift region if an external trigger pulse indicates that a physically relevant interaction has taken place

Fig. 3.20. Mean distortion parameter of tracks in the TPC of the PEP4 experiment as a function of the luminosity in the e^+e^- storage ring [PE 82, LA 83].

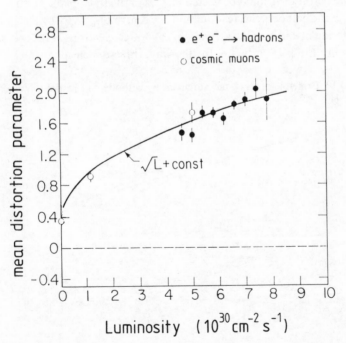

at the interaction point in the centre of the TPC [FA 79, NY 81, PE 82]. A smaller TPC chamber has been built at the TRIUMF laboratory for the investigation of rare muon decays [HA 81a]. Hexagonal end caps with 12 anode wires per sector serve for the track measurement. The gas used is an argon (80%)–methane (20%) mixture, the field strength is 150 V/cm, and the reduced field strength $E/p = 0.2$ V/Torr cm, as in the Berkeley TPC. Fig. 3.21 shows the arrangement of anode wires and segmented cathodes. A precision of 120 μm in the co-ordinate along the anode wire was obtained.

For two experiments at the electron–positron colliding storage ring LEP at CERN the TPC principle will be used for the central detector. The DELPHI experiment will use a TPC with radius $R = 1.1$ m and a length $L = 2.5$ m, while the ALEPH detector is built around a TPC with $R = 1.8$ m and $L = 4.6$ m. As a test detector for the ALEPH experiment, a chamber with a diameter of 75 cm and a length of 130 cm with a solenoidal magnet was constructed (fig. 3.22). In this chamber (TPC 90) resolutions and distortions were studied with straight ionization tracks generated by ultraviolet lasers. Spatial resolutions of 180 μm in the (r, φ)-plane and 1.5 mm in the z-co-ordinate were achieved, and the reduction of diffusion broadening was

Fig. 3.21. Segmented cathode pads and anode wires in the TPC used at TRIUMF [HA 81a].

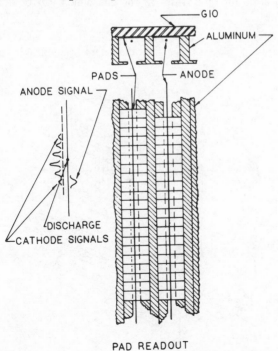

PAD READOUT

found to follow (eq. (1.76)). The properties of a gating grid between the proportional chamber and the drift region were investigated. It was shown [AM 84] that the positive ion current through the gating grid is unaffected by the presence of a magnetic field parallel to the electric field, because ions strictly follow the electric field lines due to their low $\omega\tau$ value. On the other hand, for electrons $\omega\tau$ is much larger than unity, and their motion is governed by combined electric and magnetic forces, and the opaqueness of the gated grid is reduced with increasing strength of the magnetic field.

3.6 Simulation of particle tracks by ultraviolet lasers

For the calibration of large drift chambers, in particular those discussed in sections 3.4 and 3.5, it is very useful to have straight tracks going through the chamber. Since charged particles are bent by the magnetic field in the chambers, only particles of very high momentum are practical for this purpose. An alternative way of generating straight tracks results from the observation that intense laser beams at ultraviolet wavelength produce ionization in the drift chamber gas [AN 79, BO 80b, HI 80]. The ionization density (electrons per mm^3) due to the laser beam is proportional to the square of the energy density of the laser. This is consistent with a two-photon absorption process [DE 82, RA 83, HU 85]. The ionization is due to impurities in the gas which can be completely

Fig. 3.22. Time projection chamber used for tests by the ALEPH collaboration (TPC 90). The inner diameter of the solenoid coil is 90 cm [AL 83a].

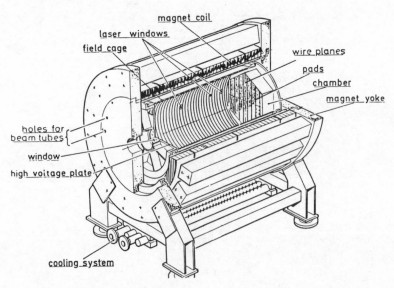

removed by appropriate filters [HU 85]. It is possible to add gaseous
additives at the ppm level in order to have controllable laser ionization
properties. For two such additives, trimethylamine and N,N,N′,N′-tetra-
methylparaphenylenediamine (TMPD), fig. 3.23 shows the measured
relation between ionization density and laser energy density at a wavelength of $\lambda = 266$ nm. The relation is again quadratic and consistent with a
two-photon absorption mechanism. Lasers of the required small beam

Fig. 3.23. Ionization density as a function of laser energy density at
$\lambda = 266$ nm for Ar/CH$_4$ (uncleaned) and doped with 5 ppm
trimethylamine and with 2 ppm TMPD. The slopes are 1.99 ± 0.06
(Ar/CH$_4$), 1.9 ± 0.1 (TMA) and 1.9 ± 0.2 (TMPD) [HU 85].

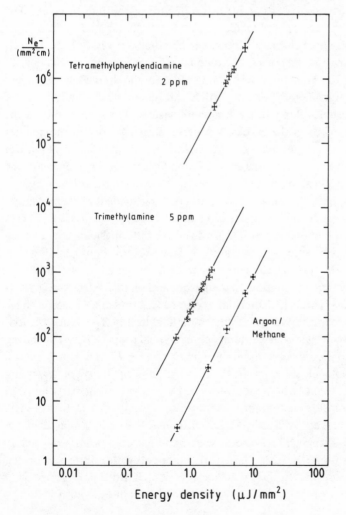

divergence are being used: nitrogen lasers ($\lambda = 337$ nm) and neodym-YAG-lasers with two frequency-doubling stages ($\lambda = 266$ nm).

3.7 Bubble chambers

In the bubble chamber [GL 52, GL 58, FR 55] a liquified gas like H_2, D_2, Ne or C_3H_8 is kept in a pressure vessel below but close to its boiling point. After the passage of ionizing radiation, the volume of the chamber is expanded by the fast movement of a piston during about 1 ms in such a way that the boiling temperature is exceeded. Along the ionizing tracks gas bubbles are forming in the liquid, due presumably to the heat developed by the recombination of ions. The growth of the bubbles is stopped when the piston moves back to its original position, restoring the initial pressure. The image of the chains of bubbles along the track is illuminated by flash lights and photographed through windows in the pressure vessel. Usually such chambers are operated in a homogenous magnetic field with a flux density up to $B = 3.5$ T. Then from the bending radius of the charged particle track the momentum $P = eBR$ of the particle can be obtained. In addition, the density of bubbles along the track can be measured. It is proportional to the energy loss dE/dx of the particle through ionization. For low momentum particles $P/mc < 3$; the mean energy loss decreases with the inverse square of the relative velocity, $1/\beta^2$ (see eq. (1.28) and fig. 1.1). A measurement of the bubble density enables therefore a determination of the particle velocity $v = \beta c$, and together with the momentum it is possible to identify the particle by obtaining a crude measurement of its mass, $m = \sqrt{(1 - \beta^2)}P/(\beta c)$.

The choice of the chamber filling is decided by the physics of the intended experiment. For studying reactions on free protons, liquid hydrogen is required. Interactions with neutrons can be investigated by measuring the difference between scattering on deuterium and on hydrogen, and correcting for nuclear effects in deuterium. If electrons, γ-rays and π^0 mesons have to be detected, a liquid with short radiation length X_0 should be used (xenon or freon). Table 10 contains some physical properties and operating conditions of liquids used in bubble chambers [BE 77, HA 81b, WE 81b].

The bubble chamber is still unique in its capability of analysing complicated events with many tracks and identifying those particles from their decays inside the chamber. An impressive example for such an event is the picture taken by the hydrogen-filled *big European bubble chamber* (BEBC) at CERN in a high-energy wide-band neutrino beam produced by 350 GeV protons (fig. 3.24). The neutrino interacts with a proton according to the equation $vp \rightarrow \mu^- D^{*+} p$, where D^{*+} is a charmed particle decaying into three charged mesons ($K^- \pi^+ \pi^+$). All details of production and decays are visible in the chamber, even a neutron from Σ^- decay is detected by the recoil proton in elastic n–p-scattering.

The use of bubble chambers as isolated detectors has diminished for the following reasons: (i) they cannot be used in colliding beam machines; (ii) at high energies (above 100 GeV) even the largest existing bubble chambers

Table 10. *Operating conditions of bubble chamber liquids*

Liquid gas	Temperature (K)	Vapour pressure (bar)	Density (g/cm³)	Expansion ratio $\Delta V/V$ (%)	Radiation length X_0 (cm)	Absorption length λ_{had} (cm)
He⁴	3.2	0.4	0.14	0.75	1027	
H¹	26	4.0	0.06	0.7	1000	887
D²	30	4.5	0.14	0.6	900	403
Ne²⁰	36	7.7	1.02	0.5	27	89
Xe¹³¹	252	26	2.3	2.5	3.9	
C₃H₈	333	21	0.43	3	110	176
CF₃Br	303	18	1.50	3	11	73
Ar	35	25	1.0	1.0	20	116

Fig. 3.24. Neutrino interaction in the hydrogen bubble chamber BEBC; picture (left) and interpretation of tracks (right) [WA 21].

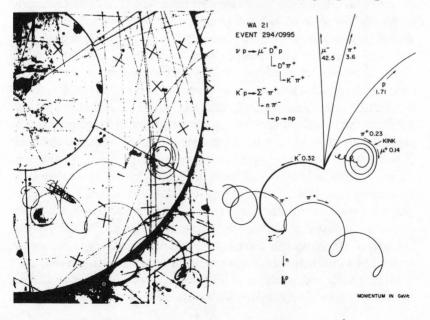

with diameters of about 5 m are not able to contain hadronic showers; a fraction of the hadron shower leaks out of the chamber, such that only a part of the primary energy of the hadron is measured in the chamber, and the calorimetric energy measurement (see section 6.2) is spoiled; (iii) the identification of muons at energies above 2 GeV requires for the absorption of pions and other hadrons a filter with a thickness of about five absorption lengths, i.e. a mass thickness of 800 g/cm^2 or a thickness of 1 m of iron – this cannot be achieved with chamber liquids; (iv) the magnetic field integral in the largest chambers amounts to $\int B \, dl \sim 10$ T m, which is not sufficient for an accurate momentum measurement. The sagitta of a particle with a momentum $P = 400$ GeV/c in a 3 m-long chamber with 3.5 T field is only $s = 0.3 \, BL^2/8P \sim 3$ mm (see eq. (7.6)).

Two of these difficulties can be solved: (ii) if a hadron calorimeter with massive plates is installed inside a bubble chamber, e.g. a liquid argon calorimeter in an argon-filled chamber [HA 82], the shower leakage can be avoided; (iii) if behind the bubble chamber an electronic 'external muon identifier' consisting of an iron absorber and proportional chambers is installed, muons can be identified: this was done for the BEBC chamber (fig. 3.25) and the 15 ft chamber at Fermilab.

Future use of bubble chambers will probably concentrate on such 'hybrid systems' of a chamber together with electronic components. One important development for the measurement of short (10^{-13} s) lifetimes are small chambers with extremely high spatial resolution, to be used as vertex detectors in conjunction with large magnetic spectrometers. Optimum resolutions are achieved with holographic readout of such small chambers.

The registration of bubble chamber tracks with the holographic technique is based on principles proposed some time ago [WE 66, WA 67, EI 79]. It enables an excellent spatial resolution and at the same time a good depth of field. The principle is illustrated in fig. 3.26 [HE 82]. A dye laser filled with Coumarin-307, pumped by an excimer (XeCl) laser producing short (10 ns) pulses at a wavelength of 308 nm, produces a parallel beam at $\lambda = 514$ nm. It is widened to the size of the bubble chamber by two lenses. The light enters the bubble chamber through a window, and the waves diffracted by the bubbles in the liquid (object ray) interfere with the original light beam (reference ray) such that they form a holographic image on the film, in this case an Agfa 10E56 emulsion on a 170 μm polyester base sucked onto metallic capstan. For reconstruction the hologram is played back by inverting the light path and using a light source at the same wavelength. In this case this light source was an argon laser ($\lambda = 514$ nm). The reconstructed image is then scanned. This 'in-line' or 'Gabor type' holography can alternatively be replaced by a two-beam geometry where one beam passes through the chamber, and the other one (the reference beam) is guided around it; interference takes place behind the chamber, where the film is situated.

An important parameter for the resolution obtainable with this method is the time interval between the passage of the particle beam and the laser flash pulse. The bubble radius r in the chamber grows with time t according to the relation

$$r = A\sqrt{t} \tag{3.5}$$

where A is a constant varying from 0.35 cm/$\sqrt{s}$ at a temperature of 48 °C to

Fig. 3.25. Top view (above) and side view (below) of the big European bubble chamber BEBC. B: beam direction; U: upper half of external muon identifier EMI; L: lower half of EMI.

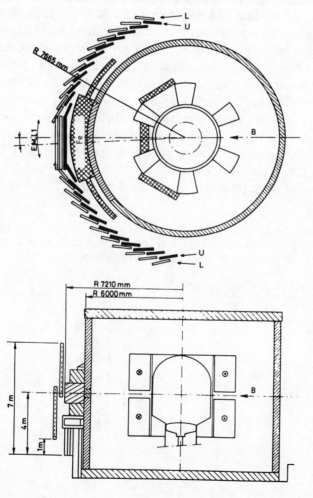

$0.023 \, \text{cm}\sqrt{s}$ at 65 °C for C_2F_5Cl and $A = 0.095 \, \text{cm}/\sqrt{s}$ at 29 K for H_2 liquid. If the delay of the laser pulse is 3 μs, at an operating temperature of 48 °C, the reconstructed bubble radius in C_2F_5Cl was 6 μm [HE 82].

As an example for the quality of images reconstructed by holographic readout, fig. 3.27 shows one of the first pictures obtained by this method in a small freon chamber (BIBC). An interaction of a 15 GeV pion in this chamber produces tracks recorded by bubbles with 8 μm diameter [DY 81, MO 80].

3.8 Streamer chambers

These detectors [CH 63] are gas-filled chambers with two planar electrodes (fig. 3.28). After the passage of a charged particle through the chamber, parallel to the electrodes, for a short time (a few nanoseconds) a very strong electric field with field strength $E > 40 \, \text{kV/cm}$ is applied to the plates. This field is perpendicular to the track direction, generating electron avalanches in the gas starting from the primary ionization along the track, with a gas amplification of more than 10^8. If the high voltage pulse to the electrodes is very short ($\sim$ ns), the discharge proceeding along the electric field lines is interrupted such that only short (0.2–1.0 mm) discharge channels ('streamers') are formed. In these discharge channels atoms are excited and emit visible light, permitting photographic pictures of the streamers. The sequence of short streamers along the ionizing particle trajectory then gives a visible track.

The process of streamer formation is depicted in fig. 3.29. The charged

Fig. 3.26. Schematic layout of the holographic optical system for the HOBC bubble chamber [HE 82] with XeCl excimer laser at $\lambda = 308$ nm. S1 to S4 are beam-defining counters.

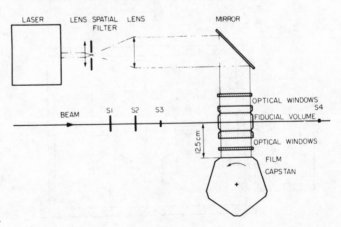

Fig. 3.27. Holographic picture of the interaction of a 15 GeV π meson in the small freon bubble chamber BIBC. The bubble size is 8 μm [DY 81].

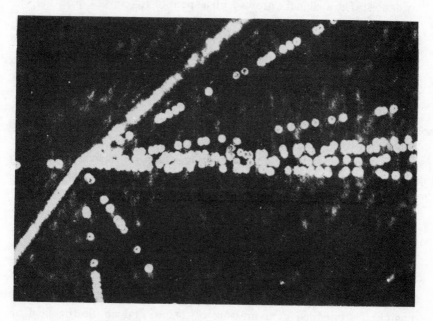

Fig. 3.28. Principle of streamer chamber (schematic).

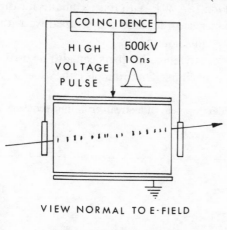

VIEW NORMAL TO E-FIELD

VIEW PARALLEL TO E-FIELD

particle track passes the chamber parallel to the electrodes. At time t_1 a cluster of ionization electrons is formed; at time t_2 this develops into an avalanche. The avalanche grows with a speed of 10^7 cm/s. A drop-shaped avalanche forms due to the different mobilities of electrons and ions (t_3). The electric field E_r due to the space charge in the avalanche adds up to the external field E. In the combined field $E + E_r$, ultraviolet photons emitted by excited atoms can produce secondary avalanches (t_4). Secondary and primary avalanches merge to form two plasma channels: one growing towards the anode, the other towards the cathode. These streamers grow with a speed of $\sim 10^8$ cm/s. If the high-voltage pulse is applied for a sufficiently long time, the streamers reach the electrodes, causing a spark.

Streamer chambers are operated with gas fillings of helium, argon and xenon or mixtures of these gases. The excellent quality of present-day streamer chamber pictures is shown in fig. 3.30. This chamber was used in an experiment investigating hadronic interactions at high energy at CERN [NA 5].

Streamer chambers with extremely good space resolution were developed at Yale for the study of charmed particle lifetimes around 10^{-13} s [DI 78]. For a particle with a mass of 2 GeV and a momentum of 20 GeV this lifetime corresponds to a mean decay path of 300 μm. The Yale chamber is operated at a pressure of 24 bar, and an electric field of 330 kV/cm is generated by voltage pulses of 0.5 ns duration. The space resolution achieved is 32 μm [SA 80]. With this technique, measurements of charmed particle lifetimes seem feasible and competitive with similar experiments using nuclear emulsions.

Unlike bubble chambers, streamer chambers can also be used as track detectors in colliding beam machines. One example is the streamer chamber

Fig. 3.29. Spatial development of a streamer in time sequence from left to right [AL 69].

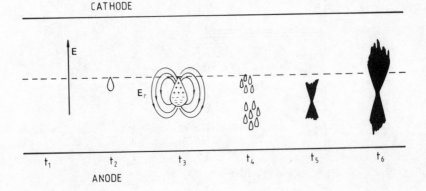

of the UA5 collaboration [UA 5] used for detecting secondary products of antiproton–proton collisions at centre-of-mass energies of 540 GeV.

3.9 Flash chambers

Another gas discharge chamber is the flash chamber developed by Conversi *et al.* [CO 55, CO 78] which is used for neutrino experiments [TA 78] and for proton decay experiments. The chamber consists of an array of rectangular tubes made of polypropylene by extrusion (fig 3.31). This planar array is placed between two metallic electrodes and filled with a neon (90%)–helium (10%) mixture. After the passage of an ionizing particle, a triggered voltage pulse is applied to the electrodes. The electric field generates a glow discharge in those cells of the array where the charged particle had passed. This discharge can be recorded by photography or by an electronic pick-up device and amplifier at the end of each cell. The flash chamber reaches a detection efficiency of 80%. The cost per area of chamber is small compared with that of drift chambers, such that it becomes possible to construct calorimetric detectors of large volume and weight (~ 1000 t), but still with a granularity of 5–10 mm. Examples can be found in the section on proton decay detectors.

Fig. 3.30. Interaction of a π^- meson at 300 GeV energy in a liquid hydrogen target. The tracks of reaction products are recorded in a streamer chamber of dimensions $200 \times 120 \times 72$ cm^3 [EC 80].

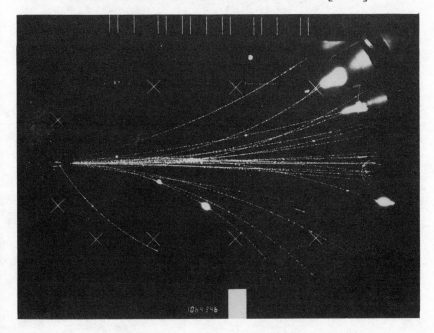

3.10 Spark chambers

Before the development of proportional and drift chambers, the spark chamber was widely used as a triggerable track detector. A set of parallel metallic plates is inserted in a volume filled with a noble gas at normal pressure, typically a helium–neon mixture. The plates are connected, in alternating order, to a pulsed high-voltage supply or to ground (fig. 3.32). If an ionizing particle has traversed the chamber, a signal generated by supplementary fast detectors (e.g. a coincidence between two scintillation counters below and above the spark chamber) triggers a high-

Fig. 3.31. Part of flash chamber made of extruded polypropylene [TA 78].

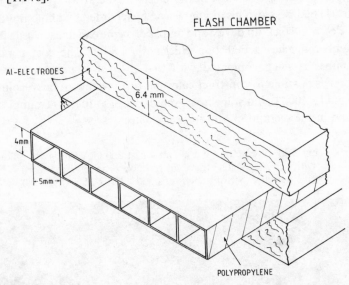

Fig. 3.32. Principle of spark chamber. PM: photomultiplier; F: pulse shaper; C: coincidence unit; V: amplifier; SG: spark gap.

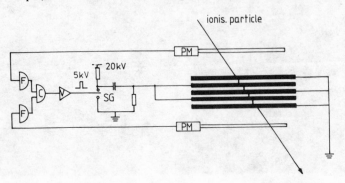

voltage pulse generator. The voltage pulse is applied to one half of the plates of the chamber [CO 55, AL 69]. The electric field between the plates is of such a strength ($E > 20$ kV/cm) that, at the place of initial ionization, avalanches and streamers are formed. The plasma channel formed in this way reaches both electrodes in about 10 ns. In general, it runs parallel to the electric field lines. The spark is photographed or registered electronically. Between discharges, the charge carriers present along the spark are sucked off by a static weak electric field (clearing field) whose direction is opposite to the pulsed field. Because of this clearing field, and of the diffusion of primary electrons, the detection efficiency of the chamber depends on the time delay between the passage of the charged particle and the application of the high-voltage pulse on the plates. Fig. 3.33 shows this dependence for a chamber filled with neon at 1.3 bar. The sensitive time of this chamber is about 10^{-6} s. If more particles cross the chamber at the same time, the

Fig. 3.33. Detection efficiency of a spark chamber as a function of the time delay between the passage of the particle and the application of the high-voltage pulse on the chamber electrodes; the parameter on the curves is the value of the voltage used for clearing the chamber after a spark [CR 60].

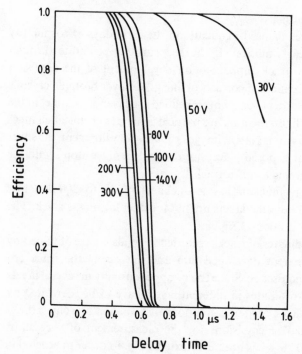

efficiency is still lower because the charge supplied by the high-voltage pulser then flows through several discharge channels.

The position of sparks is recorded optically or by magnetostrictive readout. In the latter case the electrodes are made of wires, and the discharge current flows through those few wires where the spark occurred. A magnetostrictive wire made of a Co–Ni–Fe alloy is mounted along the chamber frame orthogonal to the electrode wires. The discharge current flowing through the electrode wires near to the spark position induces a mechanical compression of the magnetostrictive wire M. This compression wave propagates along the wire M with a speed $v = \sqrt{(E/\rho)}$, where E is Young's modulus and ρ the density of the magnetostrictive wire. For $E = 2 \times 10^{11}\,N/m^2$ and $\rho = 8.1\,g/cm^3$, $v = 5\,km/s$. A coil wound around the end of M and embedded in the field of a permanent magnet picks up a voltage pulse at the time of arrival of the acoustic wave (Villari effect). This enables a measurement of the transit time of the wave from the exciting electrode wire to the pick-up coil. In this way, the position of the spark can be localized with an accuracy of about $200\,\mu m$. The intrinsic deadtime of spark chambers depends on the strength of the clearing field. It can be as low as $100\,\mu s$, but the deadtime of the high-voltage pulser is usually much larger, between 1 and 10 ms.

3.11   Nuclear emulsions

A track detector used frequently in the early days of cosmic-ray physics is the nuclear emulsion. If charged particles penetrate through photographic plates, tracks appear upon development of the emulsion [KI 10, RE 11]. The emulsion consists of fine grain silver bromide crystals with grain size of the order of $0.25\,\mu m$ embedded in a layer of gelatine. In the same way as visible light, the ionizing particle generates chemical changes in the silver bromide grains ('latent images'). During development, the silver ions of the salt in the exposed places are reduced to silver atoms, and the sequence of silver grains forms a track.

A single layer of emulsion has a thickness of 25 to $200\,\mu m$. For experiments, several hundred layers are put together forming a stack. The density of emulsion is about $3.8\,g/cm^3$.

The spatial co-ordinates of tracks recorded in emulsion are obtained by scanning the transparent developed film and measuring the track co-ordinates under a microscope. Since tracks appear usually in several layers of the stack, the co-ordinates in different layers have to be correlated by common fiducial marks. Spatial resolutions of $1\,\mu m$ can be achieved.

The high potential of emulsions for the measurement of very short lifetimes down to 10^{-13} s has been used extensively. In order to accelerate the scanning and measuring procedure, hybrid experiments use the emulsion as a target and vertex detector, and in addition have an electronic detector downstream of the emulsion in order to register and localize the most interesting events. The information of the downstream part of the detector is used to define for each event a scanning region in the emulsion. The vertex information is then combined with momenta and/or energies of particles measured in the downstream detector.

In addition to the position information on tracks, nuclear emulsion exposures can give the following quantities: the range R of the particle in emulsion, the number of delta rays formed (n_δ), the grain density g along the particle tracks, and the degree of scattering by nuclei in the emulsion. These measurements can be used to determine the energy, charge and mass of the particle. Empirical data are used to measure the range–energy relation, and the decrease of ionization energy loss with the inverse square of the particle velocity for non-relativistic particles causes a decrease of the grain density from which the velocity can be obtained. Complete identification of particles is not always possible because, for a short track, the track properties cannot be measured with sufficient accuracy.

3.12 Silicon strip detectors

The first operational device of this kind has been built by the CERN–Munich group [HY 83] for an experiment studying short-lived particles in high-energy hadronic interactions. Fig. 3.34 shows the structure of this detector. It is made of an n-doped silicon crystal wafer with a resistivity of 2 kΩ cm, a diameter of 50 mm and a thickness of 280 μm. One side of the crystal is aluminized; on the other side a sensitive area of 24 mm × 26 mm is covered with boron-implanted strips of p-type material.

Fig. 3.34. Cross-section of silicon microstrip detector with capacitive charge division [HY 83].

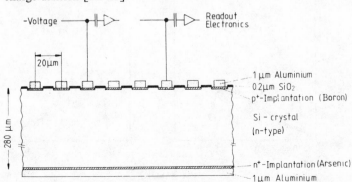

As discussed in section 2.5, these strips form p-n junction diodes if a reverse bias ($-$ 160 V in this case) is applied to the strips. A relativistic particle of charge e produces 25 000 electron–hole pairs when traversing the 280 μm of silicon. These are collected at the electrodes within 10 ns. The signals picked up at the strips measure the position of the passing particle to a precision given by the pitch of the strips. For this detector, there were 1200 strips of 12 μm × 36 mm implanted at a pitch of 20 μm. In order to reduce the number of readout channels, only every third strip (spacing 60 μm) or, alternatively, every sixth strip was read out (spacing 120 μm). The charge liberated by the particle is divided amongst the neighbouring strips by

Fig. 3.35. Position resolution of a microstrip detector as a function of voltage U and magnetic field H parallel to the strips [BE 83].

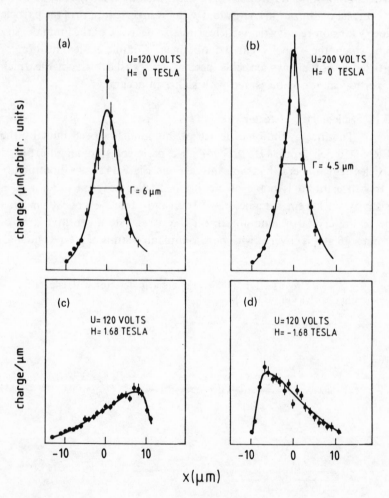

Table 11. *Properties of position detectors*

Detector	Space resolution (μm) Normal	Special	Sensitivity time (ns)	Dead-time (ms)	Direct electronic readout	Readout time (μs)	Efficiency (%)	Advantages
Proportional chamber	700	100	50	—	yes	10–10^2	100	Time resolution
Drift chamber	200	50	500	—	yes	10–10^2	100	Space resolution
Bubble chamber	100	8	10^6	10^2	no	—	100	Analysis of complex events
Streamer chamber	300	30	10^3	10	no	—	100	Analysis of multitrack events
Flash chamber	4000	2000	10^4	10^2	yes	10^3	80	Low price
Spark chamber	200	100	10^3	2–10	yes	10^4	95 (multi-track)	Simplicity
Emulsion	5	2	—	—	no	—	100	Space resolution
Silicon strip detectors	7	3	10^3	—	yes	10	100	Space resolution

capacitive coupling. The position of the particle is obtained by computing the centre of gravity of the charges collected at several readout strips. The spatial resolution obtained by this detector was 4.5 μm for the readout every 60 μm and 7 μm for 120 μm readout.

If a magnetic field of 1.68 T is applied parallel to the readout strip, the Lorentz force shifts the drifting charge carriers such that the measured co-ordinate is displaced, on average, by 10 μm [BE 83]. In addition, the spatial width of the charge distribution as obtained by the centre-of-gravity method is increased from 5 to 12 μm (fig. 3.35).

This resolution could still be improved if every strip was read out. For this case, Belau *et al.* [BE 83] calculate a resolution of 2.8 μm, of which 1 μm is due to delta electrons along the particle track. This could possibly be achieved if the electronics associated with each strip could be implanted by VLSI techniques onto the same wafer which forms the detector itself. Then the ratio of areas of the electronics compared with the detector could be unity, compared with 300 for the hybrid circuits used in the CERN–Munich detector. A first step in this direction is the integration of 60 preamplifier channels on one chip which can be connected and bonded to the detector [LU 84]. With present technology, the size of these detectors is limited to 70×70 mm^2.

3.13   Comparison of position detectors

When one compares figures of merit of position detectors, important parameters are space and time resolution and the maximum rate of data acquisition. Some of the relevant quantities have been defined in section 1.4, e.g. resolution and deadtime. The deadtime given in table 13 is the one including auxiliary equipment, like high-voltage pulsers. In addition, table 11 contains the 'sensitive time' of a triggerable detector, i.e. the time during which incoming particles are registered whether they are correlated or not with the event causing the trigger. Time overlay of different events can only be avoided if the mean time interval between events is large compared with this sensitive time.

Proportional and drift chambers are best suited for precise track recording at high event rates, while bubble and streamer chambers have the potential for the analysis of very complex events with many tracks but at a much smaller rate. The flash chamber, because of its low price and simple construction, finds increasing use in very large detectors with fine grain calorimetry, e.g. for low-rate experiments like proton decay or neutrino experiments. Silicon strip detectors and similar semiconductor detectors like charge-coupled devices (CCDs) will be used as vertex detectors in large experiments because of their unrivalled space resolution at high event rates.

4

....

Measurement of time

4.1 Photomultiplier tubes

One of the most common instruments used for measuring the time of passage of a charged particle through a detector is the photomultiplier tube (PM). Visible light from a scintillator (section 4.2) liberates electrons, by the photoelectric effect, from a thin photocathode layer at the internal surface of an evacuated glass or quartz tube. The photocathodes are semiconducting alloys containing one or more metals from the alkali group (Na, K and Cs) and materials from group V, usually antimony (Sb). For photocathodes made with two alkali components ('bialkali cathodes'). the number of photoelectrons liberated per incident photon, the quantum efficiency, reaches a maximum value of $\eta_q = 27\%$ at a wavelength of $\lambda_{max} = 380$ nm (fig. 4.1). Table 12 lists some properties of photocathode materials.

In commercial photomultiplier tubes, different geometrical arrangements of electrodes are used to collect, focus and accelerate the photoelectrons from the cathode onto a first 'dynode', an electrode made from material with a high coefficient of secondary electron emission, e.g. BeO or Mg–O–Cs. The emission of three to five secondary electrons can be achieved for one incident electron with a kinetic energy of 100 to 200 eV. For a series of 14 subsequent dynodes, maintained at potential differences of 150 to 200 V between stages, the multiplication of the number of electrons reaches 10^8. This charge of 2×10^{-11} C arrives at the anode in a time interval of 5 ns and, if the anode is connected to ground via a 50 Ω resistor, a voltage pulse of 200 mV is formed. The risetime of this pulse is about 2 ns; the total transit time inside the photomultiplier from the cathode to the anode is around 40 ns. The spread in the transit time through the photomultiplier ('time jitter') is determined by the different time spent by the photoelectrons on their way from the cathode to the first dynode. There are two effects contributing: the variation in the velocity of photoelectrons emerging from the photocathode and the different length of paths from the point of emission at the cathode to the first dynode. The spectrum of kinetic energies of photoelectrons emitted by a bialkali cathode illuminated by light of wavelengths in the interval from 400 to 430 nm extends from 0 to 1.8 eV with a peak at 1.2 eV [NA 70]. For an electric field of $E = 150$ V/cm between cathode and first dynode, the difference δ_1 in transit time between a photoelectron initially at rest and another one of kinetic energy $T_k = 1.2$ eV is $\delta_1 = \sqrt{(2mT_k)}/eE \approx 0.2$ ns. The contribution δ_2

Fig. 4.1. Spectral sensitivity N_{kr} (mA/W) and quantum efficiency η_q (%) of photocathodes as a function of wavelength λ; TU- and U-type tubes have quartz window, others have glass windows [VA 70].

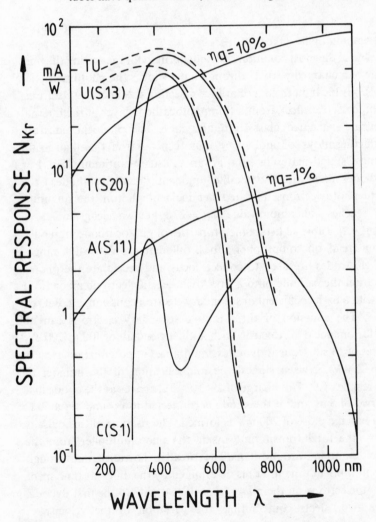

of the second effect due to geometrical path length variations depends mainly on the cathode diameter. For a diameter of 44 mm, $\delta_2 = 0.25$ ns or $\delta_2 = 0.7$ ns for the tube types XP 2020 and XP 2232B, respectively [PH 78]. This contribution seems to be the ultimate limitation in time resolution for conventional photomultipliers.

The diameters of planar photocathodes in standard commercial photomultipliers lie in the range 5–125 mm. A very large tube with a spherical cathode of diameter 508 mm has been introduced recently (Hamamatsu R 1449, figs 4.2 and 4.3). This has been developed for an experiment searching for proton decay in a large water Cherenkov counter [KU 83]. The tube has a bialkali photocathode with a maximal quantum efficiency at a wavelength of 420 nm and a gain of 10^7 for a potential difference of 2000 V

Table 12. *Properties of photocathode materials*

Material	Wavelength range (nm)	λ_{max} (nm)	Quantum efficiency $\eta_q(\lambda_{max})$	Name
AgOCs	300–1100	800	0.004	S1
BiAgOCs	170–700	420	0.068	S10
Cs_3Sb-O	160–600	390	0.19	S11
$Na_2KSb-Cs$	160–800	380	0.22	S20
K_2CsSb	170–600	380	0.27	bialkali

Fig. 4.2. Section through photomultiplier tube R 1449 with a spherical photocathode of 508 mm diameter [KU 83].

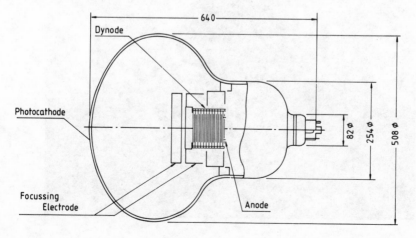

between cathode and anode. The risetime of anode pulses is $\tau_R = 18$ ns, and the variation of transit time for pulses due to one single photoelectron is $\tau_s = 7$ ns. Table 13 contains some data on time resolutions and transit time variations of some recently developed photomultipliers.

This time jitter is reduced considerably in an alternative type of electron multiplier tube, the microchannel multiplier (fig. 4.4). This device consists of an array of 10^4 to 10^7 parallel channels with a diameter of 10 to 100 μm and a length 40 to 100 times larger. The channels are holes in a lead glass plate. A suitable glass coating in the channels enables a constant potential gradient to be applied between the two ends of the channel. Electrons entering the channel at the entrance end are accelerated by this longitudinal electric field in the direction of the exit end. These electrons will strike the walls eventually, producing secondary electrons in the coating. These will also be accelerated along the channel, hit the wall somewhere downstream and produce further secondary electrons. The coating of the channel therefore acts as a continuous dynode. The multiplication factor for individual initial electrons depends on the number of times the electrons and their secondaries collide with the wall, and therefore depends on the initial orientation of the entering electron. The multiplication factor is limited by the total space charge per pulse at the exit of the channel to values below 10^7.

In order to prevent positive ions being accelerated in the inverse direction

Fig. 4.3. Photograph of photomultiplier R 1449 [KU 83].

Table 13. *Characteristic data of recently developed photomultipliers and microchannel multipliers*

	Amperex XP 2020	RCA 8854	Hamamatsu R 647–01	ITT F 4129	Hamamatsu R 1564U
Amplification	$>3 \times 10^7$	3.5×10^8	$>10^6$	1.6×10^6	5×10^5
Potential anode–cathode (V)	2200	2500	1000	2500	3400
Microchannel voltage (V)					
Risetime τ_R (ns)	1.5	3.2	2	0.35	0.27
Electron transit time τ_T (ns)	28	70	31.5	2.5	0.58
Transit time jitter (FWHM) for one photoelectron τ_s (ns)	0.51	1.55	1.2	0.20	0.09
Transit time jitter (FWHM) for many photoelectrons τ'_s (ns)	0.12		0.40	0.10	
Number of photoelectrons when measuring τ'_s	2500		100	800	
Quantum yield (%)	26	27	28	20	15
Photocathode diameter (mm)	44	144	9	18	18
Dynode material	Cu Be	GaP/BeQ			

to the electrons, the channels can be shaped in such a way that these ions hit the channel wall before they gain enough energy to produce secondary electrons themselves.

The main advantage of microchannel multipliers is the reduction of variations in transit time. Since the total transit time is only a few nanoseconds, the jitter can be as low as 0.1 ns [LE 78*b*, LE 83*a*]. Characteristic data of two commercial microchannel plate multipliers are given in table 13. The tube ITT F 4129 has a photocathode with spectral sensitivity of the S20 type (see table 12) and three microchannel plates in sequence, each of them with channels of 12 μm diameter and 500 μm length. Subsequent channels are oriented at an angle relative to the preceding one ('chevron configuration', see fig. 4.5). Between the photocathode and the first channel plate, this multiplier has a film of aluminium, 7 nm thick, in order to prevent positive ions formed in the channel penetrating the photocathode on their way back. Without such a film the lifetime of a channel plate multiplier is limited by the destruction of the photocathode by ions.

Another type of microchannel multiplier (Hamamatsu R 1564 U) consists of two plates in series behind the photocathode. As in the previous case the photocathode is protected against ions by a 13 nm-thick film of aluminium.

Fig. 4.4. Principle of microchannel multiplier [DH 77].

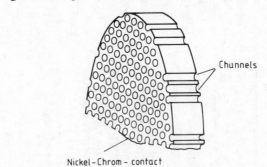

Chunnels

Nickel-Chrom-contact

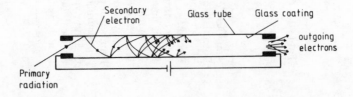

Secondary electron

Glass tube

Glass coating

outgoing electrons

Primary radiation

4.2 Scintillators

A scintillation counter has two functions: the conversion of the excitation of a transparent material caused by the ionizing particle into visible light and the transport of this light to the photocathode of a photomultiplier tube. The mechanism of scintillation [BI 64] is completely different for inorganic crystal scintillators and for organic crystalline, liquid or polymerized scintillating materials.

Inorganic scintillators are ionic crystals doped with activator ('colour') centres. The energy level diagram in such a crystal is sketched in fig. 4.6. Ionizing particles produce free electrons, free holes and electron–hole pairs (excitons). These move around in the lattice until they reach an activator centre A, which they transform into an excited state A*. A* can decay back to A with emission of light. The decay time of this scintillation light is given by the lifetime of the excited state A* and this depends on temperature T

Fig. 4.5. Schematic drawing of a two-stage microchannel multiplier in chevron configuration.

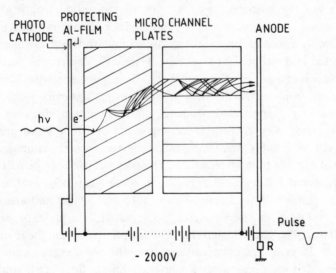

Fig. 4.6. Energy band structure in inorganic scintillating crystals.

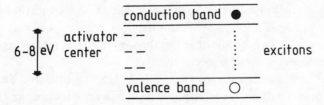

according to $\exp(-E_1/kT)$, where E_1 is the excitation energy of A*. Typical data for such scintillators are listed in table 14. The light yield in this table means the number of photons per unit of ionization energy loss of the particle detected.

Inorganic scintillators are frequently used for measurements of γ-ray and X-ray energies. Since the light yield per ionization energy is much higher than that of organic scintillators, the statistical fluctuation of the number of scintillation photons is lower and the energy resolution better than for those scintillators. This application of inorganic scintillators is discussed in section 6.

From table 14 one can infer that mean decay times of scintillation light in these inorganic crystals are usually larger than 0.2 μs. In contrast to this, decay times of light in *organic scintillators* are much shorter, in the range of nanoseconds. The mechanism of scintillation here is not a lattice effect, but proceeds through excitation of molecular levels in a primary fluorescent material which emits bands of ultraviolet light during de-excitation. This ultraviolet light is readily absorbed in most organic materials transparent in the visible wavelength region, with an absorption length of a few millimetres. The extraction of a light signal becomes possible only by introducing a second fluorescent material in which the ultraviolet light is converted into visible light ('wavelength shifter'). This second fluorescent substance is chosen in such a way that its absorption spectrum is matched to the emission spectrum of the primary fluor, and its emission should be adapted to the spectral dependence of the quantum efficiency of the photocathode. These two active components of a scintillator are either dissolved in suitable organic liquids or mixed with the monomer of a material capable of polymerization. The polymer can then be cast in any shape desired for practical application. Most frequently used are rectangular plates with thicknesses between 0.5 and 30 mm and areas up to 2×2 m². Two parameters are used to characterize the figure of merit of a scintillator: the light yield and the self-absorption length in the scintillator.

Table 15 gives the chemical structure, the wavelength of maximum emission and the decay times for several primary fluorescent substances and for two wavelength shifters [BE 71]. As a bulk material for plastic scintillators, polymeric materials of aromatic compounds (polystyrene (PST), polyvinyltoluene (PVT)) or of aliphatic ones (acrylic glasses (PMMA, 'plexiglas')) are used. The aromatic scintillators yield about twice as much light as the aliphatic ones, but the aliphatic ones are less expensive and much easier to handle mechanically.

Plastic scintillators are used frequently in large calorimetric detectors (section 6) in the shape of rectangular strips, a few metres long and with a

Table 14. *Properties of scintillating inorganic crystals*

Scintillator	NaI(Tl)	LiI(Tl)	CsI(Tl)	$Bi_4Ge_3O_{12}$	BaF_2
Density (g/cm^3)	3.67	4.06	4.51	7.13	4.9
Melting point (°C)	650	450	620		
Decay time (μs)	0.23	1.3	1.0	0.35	0.62 6×10^{-4}
λ_{max}(emission) (nm)	410	470	550	480	310 225
Light yield (photons/MeV)	4×10^4	1.4×10^4	1.1×10^4	2.8×10^3	6.5×10^3 2×10^3
Radiation length X_0 (cm)	2.59			1.12	2.1
Refractive index n	1.85			2.15	1.56
$(dE/dx)_{min}$ (MeV/cm)	4.8			8	6
Temperature coefficient of light output (%/°C)	0.22–0.9			1.7	
Radiation damage	yes			yes	
Hygroscopic	yes			no	

thickness between millimetres and a few centimetres. For this application it is important to obtain a uniform light yield over its entire length even if the scintillator light is viewed by a photomultiplier from one end only. The observed attenuation of scintillation light produced by a particle at the far end is mainly due to the absorption of the short-wavelength part of the emitted spectrum. Fig. 4.7 shows this effect for a scintillator with p-bis[2-(5-Phenyloxazolyl)]benzene (POPOP) as the wavelength shifter. In order to obtain a more uniform response it is therefore possible to filter out the part at short wavelengths by a yellow filter in front of the photocathode. The effect of such a filter cutting wavelengths below 430 nm can be seen in fig. 4.7: the light yield at the end of the scintillator near to the photocathode is reduced drastically, while the one at the far end is influenced much less.

Still, by using such a filter, the integrated light yield is diminished and, depending on experimental requirements, more value is given to a long attenuation length or maximum light yield. The scintillator on which the measurements in fig. 4.8 are based is optimized for large absorption length [KL 82b]. It is made with 3% naphthalene, 1% 2-Phenyl-5(4-biphenylyl)-1,3,4-oxadiazole (PBD) and 0.01% p-bis(o-Methylstyryl)benzene (bis-MSB) dissolved in plexiglas. Light attenuation curves for such a scintillator

Table 15. *Organic fluorescent materials and wavelength shifters*

Structure	λ_{max} emission (nm)	Decay time (ns)	Yield/ yield (NaI)
Primary fluorescent material			
Naphthalene	348	96	0.12
Anthracene	440	30	0.5
p-Terphenyl	440	5	0.25
PBD	360	1.2	
Wavelength shifter			
POPOP	420	1.6	
bis-MSB	420	1.2	

with dimensions $1800 \times 150 \times 5 \text{ mm}^3$ are shown in fig. 4.8. If a yellow filter at 430 nm is used, the attenuation length with the far end of the scintillator blackened is $\lambda_{abs} = 210$ cm. The light yield at a distance of 160 cm from the photocathode is higher by 20% compared with a similar counter made of plexiglas type 1921 (1% naphthalene, 1% PBD, 0.01% POPOP). In a large-scale production of 6000 strips of the scintillator with the bis-MSB, the frequency distribution of attenuation lengths for the individual strips was measured. Fig. 4.9 shows a standard deviation of $\sigma(\lambda_{abs}) = 18$ cm, or $\sigma(\lambda_{abs})/\langle \lambda_{abs} \rangle = 8.6\%$. This is a measure for the reproducibility of this material in large-scale manufacturing. Similar types of scintillator have been developed

Fig. 4.7. Wavelength spectrum of light produced at near end or far end of a plastic scintillator of dimensions $1800 \times 150 \times 5 \text{ mm}^3$ with and without yellow filter in front of photocathode. Scintillator is of type plexiglas 1922 [KL 82b].

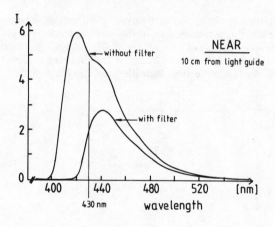

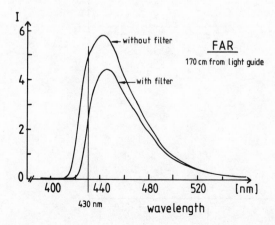

for experiments at the proton–antiproton collider at CERN [BO 81]:

(*a*) the scintillator Altustipe made with acrylic glass as base material has similar properties to the one discussed above, i.e. a light yield of 20 to 50% compared with NE 110 (polyvinyltoluene) and an attenuation length $\lambda_{abs} =$ 1.0–1.5 m;

(*b*) the type KSTI made of extruded polystyrene material; its light yield reaches 80 to 100% of the one of NE 110, the attenuation length for plates with a cross-section 3×200 mm^2 is 80 cm and the decay time is about 3 ns. Careful treatment of surfaces is required here, and machining is more delicate than for acrylic materials.

4.3   Collection of scintillation light

For the transport of scintillation light from the scintillator to the photocathode usually adiabatic light guides are used. The blue scintillation

Fig. 4.8. Light attenuation curves in a plastic scintillator of dimensions $1800 \times 150 \times 5$ mm^3. L is the distance of the ionizing track from the end of the scintillator directed towards the photocathode. ●: without filter, far end reflecting; ○: without filter, far end blackened; ▲: with filter at $\lambda = 430$ nm, end reflecting; ■: with filter, end blackened [KL 82*b*].

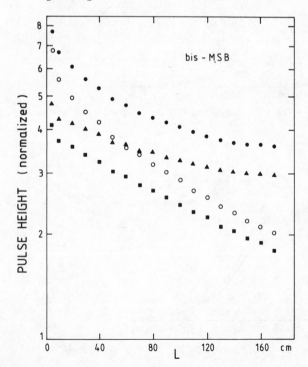

light propagates in the scintillator by multiple internal total reflections at the surfaces. For polystyrene, the refractive index is $n = 1.581$, such that light rays incident at angles larger than $\alpha_g = 39°$ relative to the normal vector of the surface are totally reflected. The front face of the scintillator plate, usually a rectangular area of area F, is imaged onto the photocathode area f by bent strips of transparent plastic material. If the minimum radius of curvature of the strips is large compared with their thickness, an angle of incidence smaller than the limiting angle α_g for total reflection can be avoided throughout the path of light. For light guides made of plexiglas, $n = 1.49$ and $\alpha_g = 42°$. Because of Liouville's theorem on the phase space of light beams, the fraction of light arriving at the photocathode is less than the ratio of surfaces, f/F.

The time resolution of a counter coupled in this way to the photocathode has two sources: the fluctuations of the transit time in the photomultiplier ('jitter', see section 4.1) and the variations of the light paths in the scintillator and the light guide. This latter contribution depends on the scintillator dimensions and for sizes exceeding 2 m it is the dominant one. This is borne out by the measurements in fig. 4.10. For long scintillators such as these the best time resolution achieved is $\sigma_t = 200$ ps.

An alternative method for light collection is based on an idea of Garwin [GA 60, SH 51]. It was implemented first for practical applications in hadron calorimeters [BA 78, SE 79]. The principle is illustrated in fig. 4.11: blue light leaves the scintillator and enters a fluorescent wavelength shifter bar through an air gap. This bar consists of plexiglas doped with a molecule

Fig. 4.9. Frequency distribution of measured light absorption length for about 6000 scintillator strips with cross-section 150×5 mm^2. The average $\langle \lambda_{abs} \rangle = 210$ cm; standard deviation $\sigma = 18$ cm.

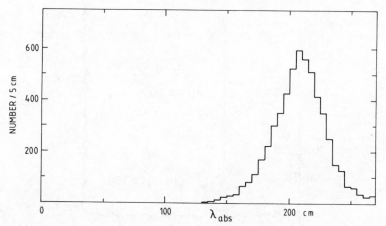

Fig. 4.10. Comparison of RMS time resolution σ_t for scintillators vs. the RMS transit time spread in the photomultipliers used. Small scintillators with dimensions below 1 cm are compared with long scintillators (length 2 m, thickness 2–5 cm, width 20–40 cm) [CA 81*b*].

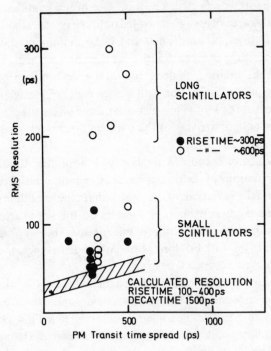

Fig. 4.11. Principle of light collection from scintillators via wavelength shifter bars.

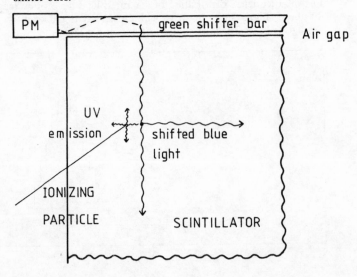

absorbing blue light and emitting green fluorescent light isotropically. A suitable agent is Benzimidazo-benzisochinolin-7-on (BBQ) with a maximum emission at $\lambda = 480$ nm. A part (10–15%) of the green light remains inside the wavelength shifter bar by total reflection and reaches a photocathode looking at one front side of the bar.

The main problems during development of this technique were (a) the search for a suitable fluorescent material adapted to the absorption of light emitted by POPOP or bis-MSB, and (b) the optimization of the self-absorption of the green light in the fluorescent bar. The solution [BA 78] consists of an admixture of 90 mg/l BBQ in plexiglas of type 218. This kind of wavelength shifter bar is now widely used in large calorimetric high-energy experiments.

Wavelength spectra of emission and absorption probabilities of BBQ are shown in fig. 4.12. The thickness D of BBQ-doped green shifter bars needed for absorption of the POPOP light from the scintillator can be inferred from fig. 4.13: for $D = 15$ mm more than 90% of the POPOP light is absorbed. The relevant absorption length for material doped with 90 mg/l amounts to $\lambda = 5.2 \pm 0.2$ mm [KL 81].

This shifter bar technique can be used to collect the light from very large scintillators with a few ($\leqslant$ four) photomultipliers. One example is the calorimetric neutrino detector of the Columbia–Fermilab–Rochester collaboration [BA 78], in which counters with an area of 3×3 m^2 are viewed by four phototubes at the corners. These four pulse heights can be used to calculate the centre of gravity of a hadronic shower of particles. Fig. 4.14 shows results of a measurement done with a $150 \times 300 \times 1.5$ cm^3 acrylic

Fig. 4.12. Absorption and emission spectra of BBQ, PBD and POPOP.

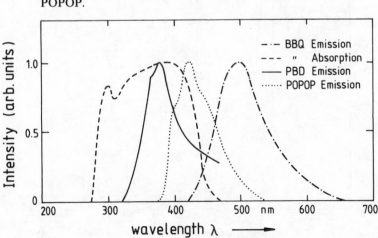

Fig. 4.13. Measurement of absorption length of POPOP light in BBQ [KL 81].

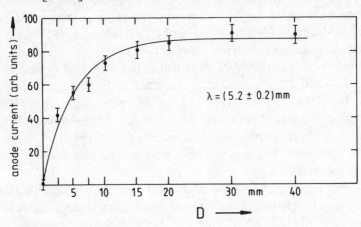

$$\lambda = (5.2 \pm 0.2)\,\text{mm}$$

Fig. 4.14. Deviation of shower position (calculated from centre of gravity of pulse heights measured at four corners of a rectangular scintillator) from real position for a shower of 100 particles; the RMS deviations are $\sigma_x = (7.3 \pm 0.1)$ cm and $\sigma_y = (7.6 \pm 0.1)$ cm; counter size $300 \times 150 \times 1.5\,\text{cm}^3$ [KL 81].

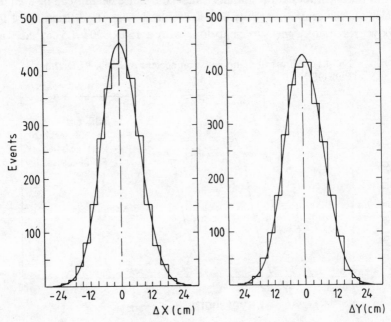

scintillator viewed in this way. The position of a shower with 100 particles can be reconstructed with an accuracy of $\sigma = 8$ cm [KL 81]. This method has the advantage that a minimal number of photomultipliers can be used. Also the construction of the shifter bars is much more straightforward than the one for adiabatic light guides. On the other hand, if several showers penetrate through a counter at the same time, their separation is not possible with this method of readout.

A further application of the wave-shifter technique is the collection of light from a lead–scintillator sandwich counter (section 6.1). For rectangular counters of this kind, a fluorescent green plate mounted on the side of a counter collects the light and is viewed by a photomultiplier behind the sandwich [HO 79]. If several green plates of different length cover two or more sides of the sandwich, the light intensity in the front and back regions of the counter can be recorded separately. This yields valuable information on the longitudinal development of an electromagnetic or hadronic shower in the sandwich counter (see section 6.1).

One disadvantage of the wavelength shifter BBQ is its decay time behaviour with two components corresponding to lifetimes of 18 ns and 620 ns, respectively, [KL 81]. This causes problems for pulse height measurements at high counting rates.

4.4 Planar spark counters

These counters consist of two planar electrodes generating an electric field above the static breakdown, i.e. at a ratio of field strength E to gas pressure p of $E/p \sim 30$–60 V/Torr cm. The primary ionization of a passing charged particle develops into a spark, and the large current drawn by this spark can be recorded as a fast rising pulse. In counters with metallic electrodes [KE 48] the spark discharges the total capacitance of the plates, leading to high temperatures and burning of the electrodes. The damaged surface gives rise to spontaneous breakdown at lower electric fields. A way of avoiding this involves using material with high resistivity ($\sigma = 10^9$–$10^{10}\,\Omega$ cm) for one of the electrodes [BA 56]. The spark then only discharges a small area of the capacitor around the primary ionization, leading to a lower energy density in the spark. Fig. 4.15 is a schematic of such a counter with copper strip readout on the semiconducting anode. The impedance of this strip line can be matched to the readout cable. Counters are operated with argon at 5–10 atmospheres, adding hydrocarbons (isobutane, ethane, 1.3-butadiene) in order to absorb ultraviolet photons from the spark, thus avoiding secondary sparking [BR 81].

The time jitter of the signal δ depends on the electric field E and the number of primary ions N as $\delta \sim 1/(E\sqrt{N})$. Measured values [BR 81] are

$\delta \sim 30\text{--}80$ ps for counters of 10×10 cm^2 area having detection efficiencies of $>95\%$. The distribution of the time difference between two parallel spark counters is not quite gaussian, showing broad tails (fig. 4.16).

Despite the excellent timing characteristics of these counters, wide application is not yet foreseeable because of the extreme difficulties in manufacturing and maintaining the high-quality surfaces. One example of

Fig. 4.15. Schematic view of typical planar spark counter [BR 81].

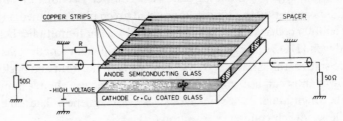

Fig. 4.16. Distribution of time difference of signals from two planar spark counters; a gaussian is fitted to the central eight bins of the experimental histogram [BR 81].

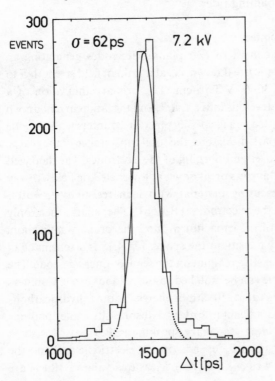

successful use of these counters is an experiment at an e^+e^- storage ring [AT 83]. The area of two counters was 9×9 cm^2, the distance between electrodes 185 μm, the resistivity of the semiconducting glass for the anode 3×10^{10} Ω cm and 6×10^{10} Ω cm, respectively. The counters were irradiated during one week with a 250 μCi γ-ray source. During this time the electrode voltage was slowly and continuously increased up to the operating voltage, such that 10^6 sparks per cm^2 were formed. By this process, the electrodes were coated with a thin film of polymerized hydrocarbon compounds. Time resolutions achieved with these two counters during the experiment at the accelerator were 76 and 148 ps, respectively. These remained constant during 150 hours of operation, corresponding to 10^7 sparks/cm^2.

5

....

Particle identification

5.1 Neutron counters

Direct neutron detection is not possible because they are uncharged and do not cause ionization in matter. Neutrons, therefore, are detected through nuclear reactions with charged secondary particles. Four main detection methods are used to count neutrons and to measure their kinetic energy:

(i) At thermal energies, nuclear reactions involving the formation of an unstable excited nucleus with delayed emission and detection of a γ-ray from the decay of this nucleus is widely used for flux measurements. The delays between formation and decay of such a nuclear state lie between 10^{-9} s and several years. Examples for neutron activation reactions are $^{63}\mathrm{Cu}(n, \gamma)^{64}\mathrm{Cu}$, $^{107}\mathrm{Ag}(n, \gamma)^{108}\mathrm{Ag}$ and $^{55}\mathrm{Mn}(n, \gamma)^{56}\mathrm{Mn}$.

(ii) At energies up to 20 MeV, prompt nuclear reactions with charged secondary products are used for energy measurements.

(iii) At energies up to 1 GeV, elastic scattering of neutrons on protons, deuterons or tritons with a detection of the recoiling target particle can be employed.

(iv) At energies above 5 GeV, the cascade of subsequent inelastic scatterings of hadrons produced by the incoming neutron can be detected in a calorimetric detector (see section 6).

Neutron detection by *prompt nuclear reactions* can proceed through the reactions

(I) $n + {}^6\mathrm{Li} \rightarrow \alpha + {}^3\mathrm{H}$

(II) $n + {}^{10}\mathrm{B} \rightarrow \alpha + {}^7\mathrm{Li}$ or

(III) $n + {}^3\mathrm{He} \rightarrow p + {}^3\mathrm{H}$

The cross-sections for these processes from the thermal region up to a few MeV are shown in fig. 5.1. The reactions are exothermic, the kinetic energy of the reaction products exceeds the one of the incident neutron by the Q-values. These are 4.76 MeV for reaction (I), 2.78 MeV for (II) and 0.77 MeV for (III).

For thermal neutrons these cross-sections are extremely large, and it is therefore sufficient to operate an ionization chamber or a proportional counter with BF_3 gas filling. The detection efficiency ε for thermal neutrons with a kinetic energy of $E_n = 0.025$ eV for a counter operating at normal gas pressure (molecular density $N = 2.69 \times 10^{19}$ cm^{-3}) is then

$$\varepsilon = 1 - \exp(-\sigma N l) \approx \sigma N l \approx 2 \times 10^2 \, l \quad \text{(for } \sigma N l \ll 1)$$

for a counter of length l cm using BF_3 with boron of the natural isotropic composition (19 % of ^{10}B).

For the detection of neutrons in the MeV range, such a BF_3 gas counter is surrounded by a layer of material with high hydrogen content in order to reduce the energy of neutrons by elastic scattering before they enter the counter ('moderation'). Such a 'long counter' is sketched in fig. 5.2. The particular geometrical arrangement is responsible for a nearly constant detection efficiency at neutron energies between 10 keV and 10 MeV, at a value of $\varepsilon \approx 0.4$ %.

Reaction (I) is the basis of the LiI crystal scintillation detector doped with europium. The ionizing α-particles and tritons from reaction (I) produce scintillation light, which is registered in a photomultiplier (see section 4.2). Such a counter can be used for the measurement of thermal neutron fluxes and of mono-energetic neutrons with energies up to 20 MeV. For neutrons of 5.3 MeV, fig. 5.3 shows a pulse height spectrum recorded in a LiI(Eu)

Fig. 5.1. Cross-sections for neutron-induced nuclear reactions as a function of neutron energy E_n (after [NE 66]).

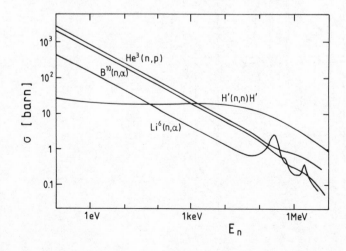

crystal of dimensions ϕ 24 mm × 2 mm. The peak at the upper end of the distribution stems from the ionizing α- and triton particles. At smaller pulse heights, thermal neutron and γ-ray background contributes to the spectrum. The relative resolution (FWHM/peak) of the peak due to α and ^{3}H-particles decreases if the crystal is cooled, from 18% at 20 °C to 10% at −142 °C.

A counter used frequently for neutrons of energies below 1 MeV is based on reaction (III). The proton and the triton from this reaction have together a kinetic energy of $E_n + 0.77$ MeV. These charged reaction products are detected in a proportional counter filled with ^{3}He and krypton at a pressure between 1 and 10 bar. As shown in fig. 5.4, mono-energetic neutrons of energy $E_n = 2.5$ MeV produce a pulse height distribution with a peak of relative width 5%. This peak is due to the ionization of the proton and triton from reaction (III). The background in this pulse height distribution

Fig. 5.2. Long counter used as neutron dose monitor in the energy region between 10 keV and 10 MeV. Neutrons enter the counter from below, are scattered in the hydrogen-rich paraffin and detected in the proportional tube P filled with BF$_3$ (after [NE 66]).

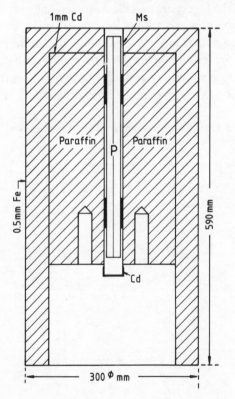

comes from the elastic scattering of neutrons on ^{3}He and from thermal neutrons.

Neutron detection by *elastic scattering* on protons, deuterons or tritons of hydrogen-rich materials makes use of the kinematic properties of the recoiling particles. If the mass of the target particle is M_r, for non-relativistic energies the recoil energy E_r in the centre-of-mass system is

$$E_r(\alpha) = \frac{2M_n M_r}{(M_n + M_r)^2}(1 - \cos\alpha)E_n$$

depending on the scattering angle α of the neutron. Using the relation $1 - \cos\alpha = 2\cos^2\theta$, for the transformation into the laboratory system, with θ now being the scattering angle in the laboratory system, one gets

$$E_r(\theta) = \frac{4M_n M_r}{(M_n + M_r)^2}\cos^2\theta E_n \tag{5.1}$$

Fig. 5.3. Pulse height spectra obtained from a LiI(Eu) scintillator exposed to neutrons of 5.3 MeV energy, at different ambient temperatures (after [MU 58]).

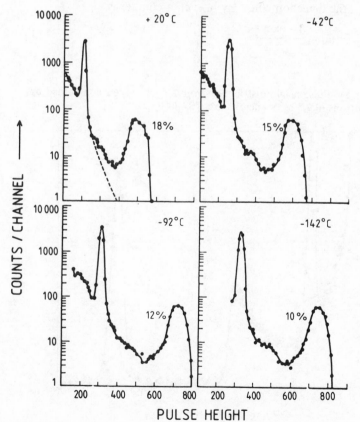

The maximum recoil energy is therefore $\approx E_n$ for n–p scattering, $0.889\,E_n$ for n–d scattering and $0.75\,E_n$ for n–t scattering. The distribution of recoil energies depends on the differential scattering cross-section as a function of angle. In the case of n–p scattering, this is isotropic, i.e. flat in $\cos\alpha$. The recoil proton energy spectrum is therefore a box spectrum extending from 0 to E_n. At neutron energies below 5 MeV, ionization chambers or proportional counters with a hydrogen-rich counting gas (e.g. CH_4) can be used for the detection of the charged recoil particles. The size of the counter should exceed the maximum proton range. At neutron energies from 1 to 100 MeV, organic scintillators, in particular liquid or plastic scintillators, are employed. The number of neutrons not detected in a counter of length L is given by $N_n(L) = N_0 \exp(-aL)$, where a contains contributions from the hydrogen and carbon contents of the scintillator:

$$a(E_n) = n_H \sigma_H(E_n) + n_C \sigma_C(E_n)$$

Here n_H and n_C are atomic densities of hydrogen and carbon, and σ_H and σ_C are the relevant elastic scattering cross-sections. Because of the relation $n_C \sigma_C \ll n_H \sigma_H$, the detection efficiency in such a counter is

$$\varepsilon(E_n) = n_H \sigma_H \frac{1 - \exp(-aL)}{a} \tag{5.2}$$

Fig. 5.4. Pulse height distribution from a ^{3}He counter bombarded by neutrons of 2.5 MeV energy (after [SA 64]).

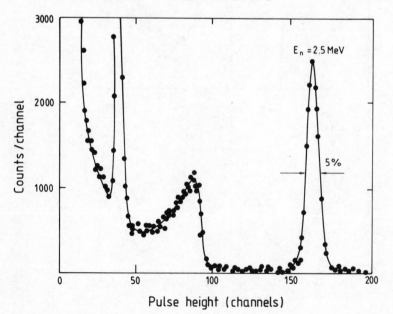

$\varepsilon(E_n)$ decreases with increasing E_n and amounts to about 20% for a plastic scintillator of thickness $L = 5$ cm at $E_n = 10$ MeV.

5.2 Time-of-flight measurement

The identification of charged particles through their flight time between two scintillation counters requires, for momenta above 1 GeV/c, very good time resolution and quite a long flight path. The time difference between two particles with masses m_1 and m_2 is, for a flight path L,

$$\Delta t = \frac{L}{\beta_1 c} - \frac{L}{\beta_2 c} = \frac{L}{c}\left[\sqrt{\left(1 + \frac{m_1^2 c^2}{P^2}\right)} - \sqrt{\left(1 + \frac{m_2^2 c^2}{P^2}\right)}\right] \qquad (5.3)$$

which for $P^2 \gg m^2 c^2$ becomes $\Delta t \sim (m_1^2 - m_2^2)Lc/2P^2$. Fig. 5.5 shows flight time differences between pairs of charged particles like electron and π meson, π meson and K meson, and K meson and proton. Using

Fig. 5.5. Differences in time-of-flight for three pairs of charged particles ($e\pi$, πK, Kp) over a flight path of 1 m.

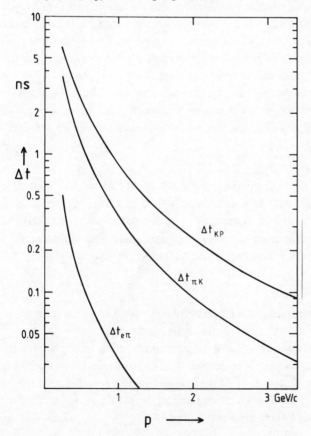

conventional scintillation counters (section 4.2) with a time resolution $\sigma_t =$ 300 ps, π–K separation at the level $4\sigma_t$ would require a flight path of 3.4 m at 1 GeV/c and 13 m at 2 GeV/c. If parallel-plate counters with a time resolution of $\sigma_t = 50$ ps could be used, the flight paths needed would be shortened to 0.6 m and 2.2 m, respectively. Since the length of the flight distance needed for particle separation at a fixed time resolution increases quadratically with the particle momentum P, the method is practical only for momenta below 2 GeV/c.

5.3 Cherenkov counters

When discussing the interaction of charged particles in matter in section 1.2, Cherenkov radiation [CH 37] was seen to be one of the manifestations of this interaction. It is electromagnetic radiation emitted by charged particles if their velocity $v = \beta c$ exceeds the light velocity c/n in the transparent medium with refractive index n traversed by the particle. The classical theory of the effect [CH 64] attributes this radiation to the asymmetric polarization of the medium in front of and behind the charged particle, representing a net electric dipole moment varying with time. In the same way as an acoustical shock wave generated by a body moving with supersonic velocity, the Cherenkov wave front can be constructed by the superposition of spherical elementary Huygens waves produced by the particle along its trajectory: during the time interval t the wave travels a distance tc/n, and the particle moves a distance $t\beta c$. From these two distances the direction of propagation of the Cherenkov wave is obtained:

$$\cos \theta_c = \frac{ct/n}{\beta ct} = \frac{1}{\beta n} \tag{5.4}$$

θ_c is the angle of the Cherenkov radiation emitted relative to the particle trajectory. This radiation is, therefore, only emitted if $\beta > 1/n$. The minimal velocity $v_s = c/n$ at which Cherenkov emission takes place is called the threshold velocity; the angle θ_c is the Cherenkov angle. The relativistic time expansion factor corresponding to the threshold velocity is

$$\gamma_s = \frac{1}{\sqrt{(1 - 1/n^2)}} \tag{5.5}$$

Fig. 5.6 shows the values of the Cherenkov angle θ_c as a function of particle velocity for different values of the refractive index n.

A more detailed consideration of the process for radiators of finite length L shows that the radiation is not only emitted at one angle θ_c, but that there is an intensity distribution around that angle θ_c caused by diffraction effects [JA 66]. This distribution has a maximum at $\theta = \theta_c$, and the distance between consecutive diffraction maxima is $\Delta\theta = (\lambda/L) \sin \theta_c$, where λ is the wavelength of the Cherenkov light. The number of photons emitted per wavelength interval and angular interval is

$$\frac{d^2N}{d\lambda\, d\cos\theta} = \frac{2\pi\alpha}{\lambda} \left(\frac{L}{\lambda}\right)^2 \left(\frac{\sin x}{x}\right)^2 \sin^2\theta \tag{5.6}$$

with $x(\theta) = [1/(n\beta) - \cos\theta]\pi L/\lambda$. In the limiting case of very long radiators $(L \gg \lambda)$, the function $(\sin x/x)^2 L/\lambda$ becomes a δ function at $x = 0$, such that the number of photons becomes, after angular integration,

$$\frac{dN}{d\lambda} = \frac{2\pi\alpha}{\lambda^2} L \sin^2\theta_c \tag{5.7}$$

The number of photons emitted in the wavelength interval from λ_1 to λ_2 is then

$$N = 2\pi\alpha L \int_{\lambda_2}^{\lambda_1} \sin^2\theta_c / \lambda^2 \, d\lambda \tag{5.8}$$

For a counter equipped with a photocathode sensitive in the visible region, $\lambda_1 = 400$ nm and $\lambda_2 = 700$ nm, such that we have

$$\frac{N}{L} = 490 \sin^2\theta_c \quad \text{photons/cm}$$

If the sensitivity is expanded into the ultraviolet region, the yield of photons can be increased by a factor of two to three. One way of achieving this goal

Fig. 5.6. Cherenkov angle θ_c as a function of the reduced particle velocity $\beta = v/c$ for a series of refractive indices n.

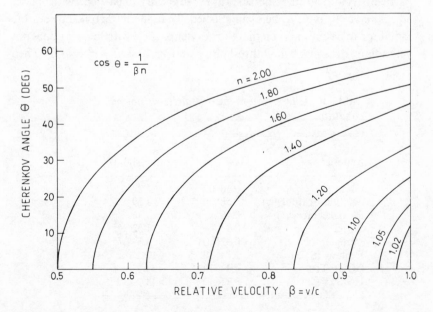

is the use of quartz windows in front of the photocathode (e.g. photomultipliers of DUVP type). Another method involves using normal glass windows, but coating them with a wavelength shifter sensitive to ultraviolet and emitting in the visible region. One such shifter is p-terphenyl, applied as a 0.2 mg/cm^2 layer on the glass window and protected by a 250 Å-thick layer of MgF_2 [GA 72]. The performance of this tube is at least the equal of the one obtained from quartz face phototubes.

Common radiator materials for Cherenkov counters are listed in table 16 with their refractive indices and threshold dilatation factors γ_s.

Between the highest refractive indices attainable with gases (e.g. pentane) and the lowest values for transparent solids or liquids there is a gap, which could only recently be closed with the development of silica-aerogels. The molecular structure of these substances contains m times (SiO_2) and $2m$ times (H_2O), where m is a natural number. They have a refractive index between $n = 1.025$ and 1.075 such that Cherenkov thresholds in the region of γ_s from 3 to 5 become accessible. The production of silica-aerogel with $n = 1.03$ to 1.05 in blocks of $18 \times 18 \times 3$ cm^3 is possible [HE 81]. The Cherenkov light in this case is collected with cylindrical mirrors [CA 81a] or with a hollow box behind the radiator covered internally with diffusing material [AR 81]. With five photomultipliers each having a 110 mm-diameter photocathode, 6 to 12 photoelectrons were collected from a radiator of 18 cm thickness (fig. 5.7).

The length of Cherenkov threshold detectors needed for separation of particles of momentum P increases as P^2: suppose two particles with masses m_1 and $m_2 > m_1$ have to be distinguished. Then the refractive index of the radiator can be chosen such that the heavier particle with mass m_2 does not yet radiate, or is just below threshold, $\beta_2^2 \approx 1/n^2$, and $n^2 = \gamma_2^2/(\gamma_2^2 - 1)$. Then

Table 16. *Cherenkov radiators, gases at normal conditions (STP)*

Material	$n - 1$	γ_s (threshold)
Glass	0.46–0.75	1.22–1.37
Scintillator (toluene)	0.58	1.29
Plexiglas (acrylic)	0.48	1.36
Water	0.33	1.52
Aerogel	0.025–0.075	4.5–2.7
Pentane (STP)	1.7×10^{-3}	17.2
CO_2 (STP)	4.3×10^{-4}	34.1
He (STP)	3.3×10^{-5}	123

the amount of Cherenkov light from the particle with mass m_1 is proportional to

$$\sin^2 \theta_c = 1 - \frac{1}{\beta_1^2 n^2}$$

which for $\gamma \gg 1$ becomes $\quad \beta \gg 1 \qquad P = m v t$

$$\sin^2 \theta_c \approx \frac{c^2(m_2^2 - m_1^2)}{P^2} \tag{5.9}$$

In a radiator of length L, detecting photons with a quantum efficiency of 20%, the number of photoelectrons is

$$N_p = \frac{100 \, Lc^2(m_2^2 - m_1^2)}{P^2 L_0} \tag{5.10}$$

Fig. 5.7. Cherenkov threshold counter with silica-aerogel as radiator; particles enter the counter from the left side, the Cherenkov light is reflected by a mirror (broken line) towards the five photocathodes with multipliers (PM) [AR 81].

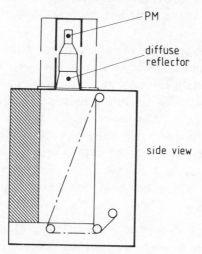

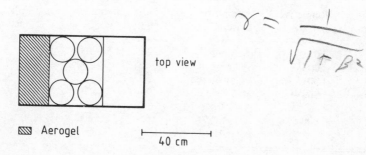

$\gamma = \dfrac{1}{\sqrt{1 - \beta^2}}$

where $L_0 = 1$ cm. In order to obtain $N_p = 10$ photoelectrons, a length

$$\frac{L}{L_0} = \frac{P^2}{10(m_2^2 - m_1^2)c^2} \tag{5.11}$$

is required in the optimistic case, assuming that a radiator with exactly the refractive index required above can be found.

For practical purposes one can use a combination of threshold counters with different refractive indices, as indicated in table 17. By using two or more of these counters, pions, kaons and protons can be identified in the momentum ranges given in fig. 5.8.

Apart from this utilization of the Cherenkov threshold, the *angle* of Cherenkov emission can also be measured in order to identify particles. The conical emission pattern around the radiating particle can be focused into a ring-shaped image. An adjustable diaphragm at the focus transmits Cherenkov light emitted in a small angular range into a phototube. Changing the radius of the diaphragm allows a scan through regions of velocity. Differential gas Cherenkov counters [LI 73] correcting for chromatic dispersion in the radiator (DISC) have achieved velocity resolutions of $\Delta\beta/\beta \approx 10^{-7}$.

The length of such counters is limited to a few metres. Separation of two kinds of charged particles with different masses, m_1 and m_2, with a Cherenkov counter of fixed length L is only possible up to a maximum momentum of the particles; for threshold counters this limiting momentum P_{max} can be inferred from eq. (5.9). The relevant figure is the number of photons N emitted by the faster one of the two particles, with mass m_1, if the slower one having mass m_2 does not radiate at that time, i.e. it is just below the Cherenkov threshold. Because $N/L = 490 \sin^2 \theta_c$ photons/cm, this number is determined by the Cherenkov angle for the faster particle.

Table 17. *A selection of possible Cherenkov threshold counters* [*LE 81c*]

Counter	Refractive index n	Radiator material	Length of radiator (cm)	Length of counter (cm)	Yield (photo-electrons)
A	1.022	Aerogel	20	50–100	5–6
B	1.005	? (Aerogel)	?	50–100	?
C	1.001 77	Neopentane	30	50	≈ 10
D	1.000 49	(N_2O–CO_2) or freon 14	100	≈ 120	≈ 10
E	1.000 135	(Ar–Ne) or H_2	185	≈ 200	≈ 5

According to eq. (5.9) $\sin \theta_c$ drops linearly with $1/P$, and for π–K separation the angle is $\theta_c = 48$ mrad at $P = 10$ GeV/c. The corresponding curve for threshold counters is shown in fig. 5.9. The maximum momentum at which π–K separation is possible can then be read off the graph if the number of photoelectrons required for a signal and the length of the counter are known, giving the minimum Cherenkov angle required for detection. For $N_p \geqslant 10$ and $L = 100$ cm, $\sin^2 \theta_c$ has to be larger than 0.001 or $\theta_c \geqslant 30$ mrad. From fig. 5.9, $P_{\max} \sim 16$ GeV/c for threshold counters separating π and K mesons. For other pairs (m_1, m_2) of particles to be separated, the maximum momentum can be inferred from the lower part of the figure, where the scale is changed by a factor $\sqrt{(m_2^2 - m_1^2)}$.

The curves corresponding to differential counters and DISC counters are also shown in fig. 5.9. The resolution of these instruments is superior to the simple threshold counter. Momentum limits for π–K separation are higher by a factor of 10 for differential counters and by a factor of 30 to 40 for DISC counters. With these most sophisticated instruments π–K separation is still possible at a momentum of 500 GeV/c.

A velocity spectrum of charged particles in a short secondary beam

Fig. 5.8. Momentum ranges in which the three particles, π meson, K meson and proton, can be positively identified by a combination of two or more threshold Cherenkov counters listed in table 17 [LE 81c]. Below, ranges for identification by time-of-flight measurement are indicated: for an RMS time resolution of $\sigma_t = 0.15$ ns, a $3\sigma_t$ separation between particle pairs is required; flight path 3 m (full line) or 7 m (dashed line).

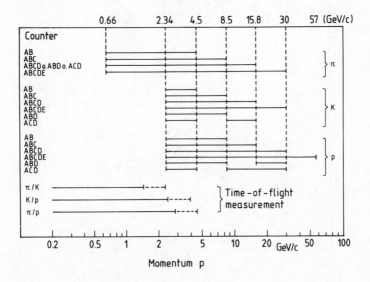

produced by an external proton beam of momentum 24 GeV/c is shown in fig. 5.10. The secondary beam was tuned to 15 GeV/c particles, and peaks corresponding to Σ^- and Ξ^- hyperons are visible, apart from the signals from mesons and antiprotons.

The differential Cherenkov counters considered above had a diaphragm of variable radius in order to scan the velocity distribution of a beam. Alternatively, the diaphragm can be fixed and the gas pressure varied to achieve the same scan. Such a counter [BO 83] is sketched in fig. 5.11. The counter was used in a momentum-selected secondary beam of positive particles produced in the forward direction by protons of energy 400 GeV hitting a target. The counting rate as a function of gas pressure shows three

Fig. 5.9. Cherenkov angle at maximum momentum for π–K separation by threshold Cherenkov counters, differential and DISC–Cherenkov counters (after [LI 73]).

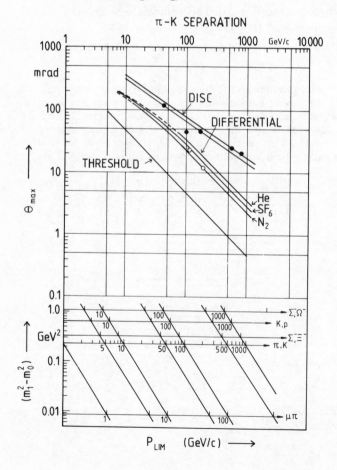

Fig. 5.10. Velocity distribution of charged hyperons with 15 GeV/c momentum in a short secondary beam [LI 73].

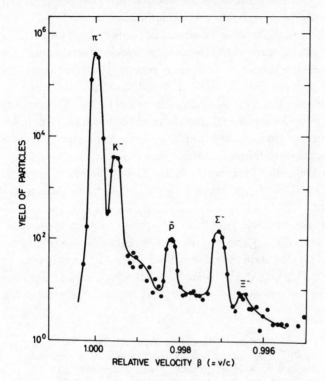

Fig. 5.11. Differential Cherenkov counter with fixed diaphragm and change of refractive index by pressure variation [BO 83].

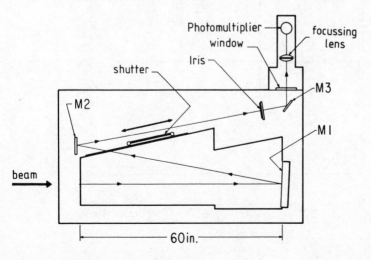

peaks corresponding to π and K mesons and protons (fig. 5.12). Since the two kinds of mesons decay into neutrinos, such counters have been used to measure indirectly fluxes of neutrinos and antineutrinos.

While the differential Cherenkov counters can only be used for particles parallel to the optical axis of the detector, a velocity measurement for diverging particles from an interaction region requires a different approach. Seguinot and Ypsilantis [SE 77] have proposed the idea of a Cherenkov ring imaging detector (fig. 5.13). A spherical mirror of radius R_M centred at the interaction point focuses the Cherenkov cone produced in the radiator between the sphere of radius R_D and the mirror into a ring-shaped image on the detector sphere of radius R_D. Usually $R_D = R_M/2$.

Since the focal length of the mirror is $R_M/2$, the Cherenkov cones of opening angle $\theta_c = \cos^{-1}(1/\beta n)$ emitted along the particle's path in the radiator are focused onto a ring with radius r on the detector sphere. For $R_D = R_M/2$, the opening angle θ_D of this ring equals θ_c, in first approximation. The radius r of the ring image gives the Cherenkov angle via $\tan \theta_c = 2r/R$, and from this we obtain the velocity $\beta = 1/n \cos \theta_c$. The relative error on β is $\Delta\beta/\beta = \sqrt{[\tan^2 \theta_c (\Delta\theta_c)^2 + (\Delta n/n)^2]}$. Neglecting the error from the uncertainty in n, one obtains $\Delta\gamma/\gamma = \gamma^2\beta^3 n \sin \theta_c \, \Delta\theta_c$, and the error in the

Fig. 5.12. Counting rate of Cherenkov counter in fig. 5.11 as a function of gas pressure. The counter is traversed by particles in a secondary beam produced by 400 GeV protons. Maxima correspond to π^+ mesons, K^+ mesons and protons selected to have 165 GeV/c momentum [BO 83].

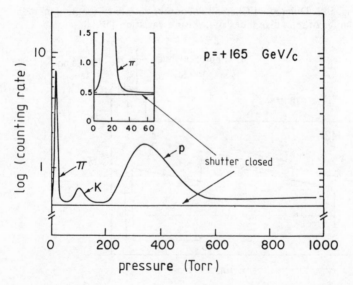

momentum $P = m\beta\gamma$ of the particle detected is [YP 81]

$$\frac{\Delta P}{P} = \frac{\Delta\gamma}{\gamma\beta^2} \qquad (5.12)$$

The critical point of such a RICH counter is the development of photoionization detectors. At present [EK 81], proportional chambers with an admixture of photosensitive triethylamine (TEA) are under study. Such a photon detector is shown in fig. 5.14. Behind the ultraviolet transparent CaF_2 window there are three gaps and a proportional chamber (PC). In gap C photons are converted by TEA to photoelectrons, gap PA serves for preamplification, gap T for transfer and PC for avalanche multiplication. Three photoelectrons per incident 10 GeV/c pion traversing 1 m of argon Cherenkov radiator at 1.2 atm pressure have been observed. Development of such detectors is continuing. Amongst the possible improvements is the study of a new photosensitive gas [NA 72], tetrakis-dimethylaminoethylene (TMAE), having a photoionization potential of 5.4 eV, lower than that of TEA which is 7.5 eV. Using this gas, evidence for Cherenkov ring imaging has been obtained as shown in fig. 5.15, where ten events have been overlapped in the picture [SA 81].

For one of the four experiments at the large electron–positron collider

Fig. 5.13. Principle of ring imaging Cherenkov counter (RICH) [SE 77].

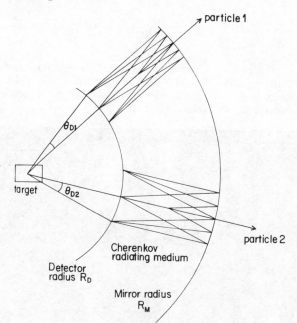

(LEP) at CERN, RICH counters form an essential part of the apparatus; they are designed to enable the identification of π and K mesons in a jet of hadrons formed in the e^+e^- interaction.

5.4 Transition radiation detectors

If one examines the measured curve for the intensity of Cherenkov emission as a function of the gas pressure in fig. 5.12, there is evidence for emission of light even at vanishing pressure in the Cherenkov counter, i.e. below the Cherenkov threshold [BO 83]. This effect can be explained as

Fig. 5.14. Photon detector for ring imaging Cherenkov counter with CaF_2 window and four planar chamber gaps: C, for conversion of ultraviolet light to electrons; PA, for amplification; T, for electron drift; and PC, for proportional chamber. Dimensions in millimetres [EK 81].

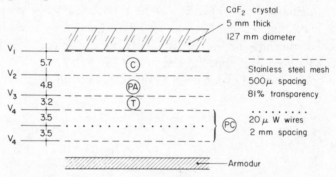

Fig. 5.15. Evidence for Cherenkov ring image; the photons are detected in a multistep spark chamber with the photosensitive gas TMAE. Photons from ten charged particles are overlapped in the picture; the central spot is due to beam particles [SA 81].

radiation emitted when the particles traverse the boundary between the dense medium outside and the vacuum inside the counter and again the boundary between vacuum and denser medium at the exit. This is the 'transition radiation' (TR) mentioned in section 1.2. It is always emitted if a charged particle traverses a medium with varying dielectric constant, e.g. a periodic series of foils and air gaps. The radiation is emitted from the interfaces between the two materials. One can imagine its generation in the following way: the charged particle moving in the vacuum forms, together with its image charge in the denser medium, an electric dipole; the electric field strength of this dipole varies while the particle approaches the boundary plane and vanishes upon its entry into the denser medium. This dipole varying in time produces radiation. For a relativistic particle with the time dilatation factor $\gamma = E/mc^2$, the radiation is concentrated on a cone with opening angle $1/\gamma$. As calculated by Ginzburg and Frank [GI 46], and as shown in section 1.2.1, the intensity of TR increases linearly with γ, and has indeed a sharp maximum on a cone with a half opening angle $\phi \sim 1/\gamma$. If a periodic arrangement of many foils and gaps is used as a radiator, interference effects [AR 75, FA 75] will produce a threshold effect in γ. Counters measuring this TR can therefore be used in order to discriminate between particles of different mass but of same momentum by means of their γ factor.

Practical applications have followed the demonstration by Garibian [GA 73] that TR is emitted also in the X-ray region. Actual counters consist of a radiator followed by a proportional chamber for the detection of the X-rays emitted forward. Since the absorption of X-rays in the radiator material behaves as $Z^{3.5}$, the atomic number of the foils has to be as low as possible. In the pioneering work of Willis, Fabjan and co-workers [CO 77] the technology of thin lithium ($Z = 3$) foils has been mastered. As a counting gas for the X-ray detector, xenon ($Z = 54$) has been used.

The pulse height spectrum in such a xenon chamber behind a radiator consisting of 1000 lithium foils of 51 μm thickness is shown in fig. 5.16 for incident electrons of momentum 1.4 GeV/c ($\gamma \sim 2740$). This curve is labelled 'e$^-$ with Li'. This pulse height comes from TR photons as well as from the electron traversing the xenon chamber. The impact points of the electron and the photons are very close in the chamber such that their separation is not possible here. In addition, fig. 5.16 shows the pulse height spectrum from π mesons of 1.4 GeV/c momentum corresponding to the energy loss by ionization with the characteristic tail of the Landau distribution towards high pulse heights ('π with Li'). The third histogram in the figure ('e$^-$ with dummy') shows that electrons of the same momentum do not produce TR in a radiator without variation of the dielectric constant but with the same mass as the original radiator.

The increase of the total radiated TR energy with γ is mainly due to an increase in the average energy of TR quanta, as shown in fig. 5.17. Here, results of measurements with electrons producing TR in different kinds of lithium foil radiators are given. The detector for the TR X-rays is in all cases a proportional chamber filled with Xe/CO_2 in the ratio $80\%:20\%$. The increase of the average TR quantum energy with electron momentum P_e depends on the arrangement of foils in the radiator. In the range covered by this experiment, the γ factor varies between 1000 and 6000. From these experiments we can conclude that (i) TR detectors at the moment can be used for $\gamma > 1000$, i.e. for electrons above $0.5\,GeV/c$ and pions above $140\,GeV/c$; (ii) the extension of this method below $\gamma = 1000$ requires the detection of 1–5 keV X-rays.

When one separates a TR signal from the pulses caused by the ionization of a non-radiating particle, the upper end of the Landau distribution for the energy loss is disturbing because it overlaps with the region of TR pulse heights (see fig. 5.16).

Recently, Ludlam *et al.* [LU 81] have shown that an improvement in the separation of particles can be obtained by not only measuring the total energy deposited by TR quanta but counting ionization clusters along the

Fig. 5.16. Pulse height spectrum from a xenon filled proportional chamber of 1.04 cm thickness behind a transition radiator (1000 Li foils of 51 μm thickness) exposed to a beam of electrons and π mesons of 1.4 GeV/c momentum [FA 80].

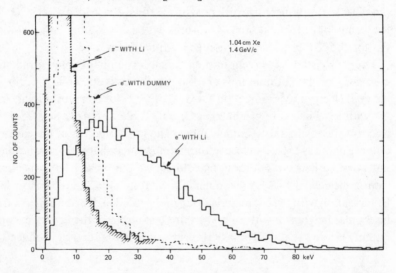

track. The number of such clusters from an ionization particle track obeys a Poisson distribution, while the upper end of an energy loss curve has a very long tail. If therefore a charged particle below transition threshold has to be separated from a particle with TR, the region of overlap becomes much smaller for the cluster counting method. Fig. 5.18 shows the principle of method [LU 81], and fig. 5.19 the distribution in cluster number N and in deposited charge Q for 15 GeV pions and electrons [FA 81]. Also shown is the pion rejection against electron efficiency for a particular cluster threshold energy of 4 keV, obtained with a detector of 12 sets of 35 μm lithium foils, each one followed by a xenon proportional chamber. The detector has a total length of 66 cm and a thickness of 0.04 radiation lengths. For a 90% electron efficiency, a pion rejection of 8×10^{-4} is obtained. The corresponding figure for a radiator made of pure carbon fibres of 7 μm thickness is 2×10^{-3}. With a similar detector of 132 cm length, a kaon rejection of 10^{-2} has been achieved for a 90% pion detection efficiency, as shown in the points labelled 'N(Exp.A)' in fig. 5.20. Also shown are measurements of Commichau *et al.* [CO 80] using *only* charge measurements, labelled 'Q(Exp.B)'. This detector nearly reaches the rejection obtained with the cluster method in Exp.A.

Fig. 5.17. Mean energy of TR X-ray quanta in a TR xenon/CO_2 detector as a function of electron momentum P_e. Measurements for different radiators are displayed [CO 77].

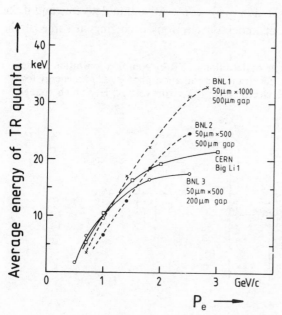

5.5 Multiple ionization measurement

If we consider the momentum range in which the particle identification methods treated so far are applicable, time-of-flight can be used at low momenta up to 2 GeV/c, Cherenkov counters up to about 20 GeV/c, DISC Cherenkov counters up to 200 GeV/c and TR counters above a value $\gamma = P/mc > 1000$. For π mesons these momentum limits correspond to time dilatation factors $\gamma = 14$, 140 and 1400, respectively. Since DISC counters can only be employed for parallel beams of particles, there is a gap in the range of γ between 100 and 1000. This gap can be closed by the ring imaging Cherenkov counters (RICH, section 5.3) or by detectors measuring the *relativistic rise* of the specific energy loss dE/dx by ionization, as shown in figs 5.21 and 1.1. This rise amounts to 50% of the minimal value $(dE/dx)_0$ at $\gamma \sim 4$, in gases, and a determination of γ through this increase requires a very precise measurement of the mean energy loss (of the order of 2 to 5%). In the momentum range below 1 GeV/c, the steep decrease of dE/dx with $1/\beta^2$ can also be used for particle identification, as an alternative to the time-of-flight measurements.

The statistical distribution of energy losses in a thin layer of gas is shown in fig. 1.2. Because of the asymmetric tail at the upper end of the distribution, the 'Landau tail', the statistical precision in determining the mean of the distribution does not increase in the same way as for a gaussian distribution if the thickness of the gas layer traversed and consequently the number of charges liberated increases. However, the resolution increases if the energy loss is measured in many consecutive thin detectors and if the large pulse heights from knock-on electrons occurring in some of the

Fig. 5.18. Principle of detection of TR by counting ionization clusters along the track. TR: transition radiation photon; dE/dx: energy loss by ionization; E: electric field; D.V.: drift voltage; H.V.: high voltage [LU 81].

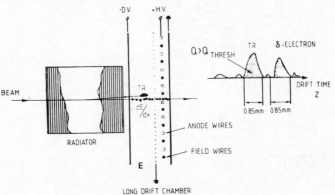

detectors are removed. This is done by taking the mean of the lowest 40–60% of measured ionization values. This sampling method with truncation reduces fluctuations in the mean and permits a measurement of energy loss precise enough in order to distinguish particles if their

Fig. 5.19. Results on pion–electron separation by TR at 15 GeV/c momentum. Upper part: differential distributions in the number N_{cl} of clusters and the charge Q deposited in one chamber. Lower part: suppression factor for π mesons ('pion contamination') vs. electron detection efficiency for three methods of discrimination. Q: cut in charge distribution; N(ADC) and N(DISC): cut in cluster number by different electronic methods [FA 81].

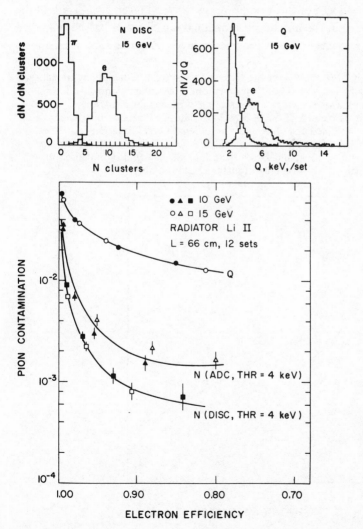

momentum is known. As can be seen from fig. 5.21, the ratio of mean energy losses of π and K mesons at a momentum of 100 GeV/c is only about $I_\pi/I_K \sim 1.05$. As an illustration, the difference $I_\pi - I_K$ divided by I_K is extracted from fig. 5.21 and shown in fig. 5.22, as a function of momentum P. At $P = 100$ GeV/c, with $(I_\pi - I_K)/I_K = 0.05$, π–K separation requires a measurement of dE/dx with an RMS resolution of better than 2%. Such a resolution can be achieved by using more than one hundred proportional chambers with a total layer thickness L of several metres of gas. With a detector consisting of 128 chambers, by using the 51 smallest pulse heights, a resolution of $\sigma_r = \sigma(\mathrm{d}E/\mathrm{d}x)/(\mathrm{d}E/\mathrm{d}x) = 2.5\%$ has been obtained for π mesons and protons of 50 GeV/c momentum [LE 78a].

The dependence of this resolution on the total layer thickness L of the detector and on the number of individual chambers N has been studied [AD 74]. The result is that for a fixed layer thickness L/N of chambers

Fig. 5.20. Results on the separation of π and K mesons by transition radiation at 140 GeV/c momentum. The suppression factor for π mesons is shown vs. the detection efficiency for K mesons. Experiment A [FA 81] uses 24 radiators of carbon fibres, each with a xenon filled chamber, and a total length of 132 cm. Experiment B [CO 80] has 20 radiators made of 5 μm-thick mylar foils with one chamber each, and a total length of 147 cm. Q: discrimination using charge; N: discrimination using cluster number.

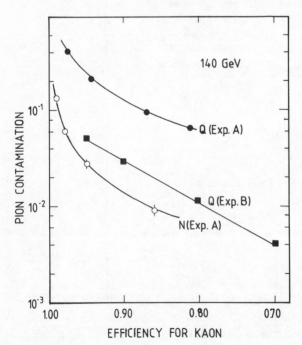

Fig. 5.21. Mean ionization energy loss in a 1 cm-thick layer of 80% argon and 20% methane at standard conditions for five kinds of charged particles [MA 78].

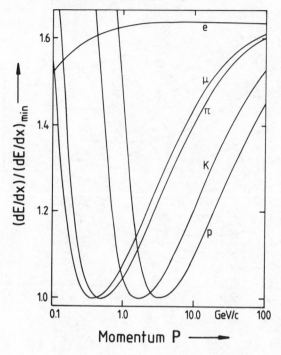

Fig. 5.22. Relative difference between the mean energy losses of π and K mesons, $|I_\pi - I_K|/I_K$, as a function of particle momentum.

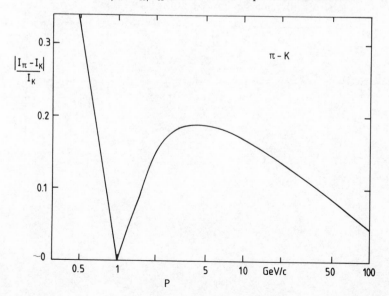

Fig. 5.23. Experimental resolution of energy-loss measurements (FWHM/mean) for N gas counters of thickness T. $L=NT$ is the total detector length; the gas filling is argon at STP [AD 74].

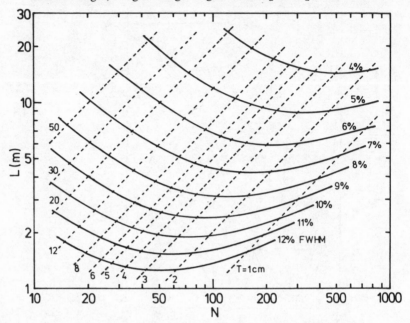

Fig. 5.24. Experimental resolutions in energy-loss measurements achieved in large detectors as a function of total detector length L multiplied with teh gas pressure p [LE 83b].

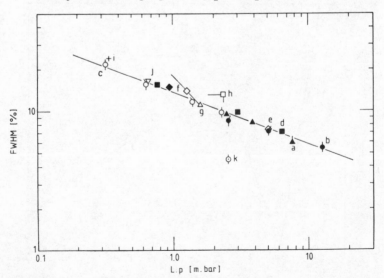

('sampling thickness') the relative resolution σ_r of the truncated mean decreases approximately with $1/\sqrt{N}$ and with $1/\sqrt{L}$, as expected on a statistical basis. If the total thickness L is kept constant, however, the resolution is not independent of the number of measurements N, but there is an optimum number of chambers N^*, and therefore an optimum thickness of chambers $T^* = L/N^*$. A parametrization of these results is displayed in fig. 5.23. For a detector with a total thickness of $L = 4$ m the optimum number of chambers is, according to this figure, about $N^* = 150$ and the optimum sampling thickness about $T^* \approx 3$ cm, for argon at standard conditions.

Concerning the dependence of the resolution σ_r on the gas pressure p, in the first approximation the statistical fluctuation of the truncated mean

Fig. 5.25. Relativistic rise of mean energy loss by ionization in argon–methane mixtures at different gas pressures from 0.25 to 7.13 bar [LE 81*b*].

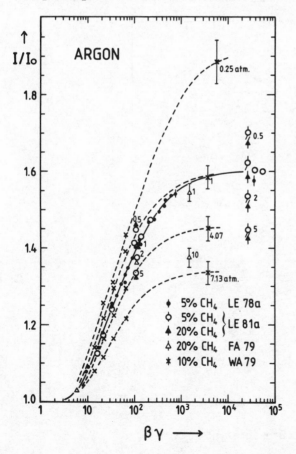

should decrease with the number of charge carriers liberated in the ionization process, and therefore $\sigma_r \propto 1/\sqrt{p}$. Together with the $1/\sqrt{L}$ dependence discussed above this leads to a scaling law which reads

$$\sigma_r \propto \frac{6.2}{\sqrt{(Lp)}} \quad \% \tag{5.13}$$

for a fixed sampling thickness of 0.03 m argon, measuring L in metres and p in bar. Another form of this relation is obtained for a constant number of chambers N and variable chamber thickness $T = L/N$:

$$\sigma_r \propto \frac{6.2}{\sqrt{(NTp)}} \quad \%$$

Fig. 5.26. Relative difference D between mean energy losses of two kinds of particles (π/p and e/π), divided by the relative resolution σ, as a function of gas pressure. The particle momentum is 15 GeV/c, the energy loss is measured in 64 counters of 4 cm thickness [LE 81a].

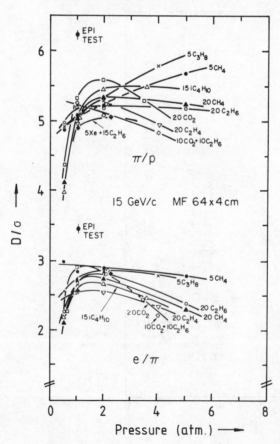

In large experiments, the improvement of resolution with increasing gas pressure expected from eq. (5.13) was not completely confirmed; fig. 5.24 shows a compilation of results obtained in big operational detectors [LE 83b] in which the gas pressure, or alternatively the total detector thickness L, was varied. In these experiments, the resolution decreases according to the relation

$$\sigma_r = 5.8\% \cdot (Lp)^{-0.37} \tag{5.14}$$

where L is measured in metres and p in bar. In addition to this result, particle identification with gas counters at higher pressures is impeded by the 'density effect' discussed in section 1.2.1, eq. (1.23): the relativistic rise in dE/dx is smaller at higher density. For argon, the increase is only 30% at 7 bar pressure compared with 55% at 1 bar (see fig. 5.25). Together with the deviations from the scaling law, this behaviour makes particle identification at 1 bar nearly as efficient as at higher pressure. Fig. 5.26 shows the relative differences of truncated means, $D = (I_\pi - I_p)/I_p$ and $D = (I_e - I_\pi)/I_\pi$, divided by the resolution σ_r, for different gas pressures and different gas mixtures. For all measured gas mixtures, D/σ_r increases when increasing p from 1 to 2 bar. For most gases, the curves for D/σ_r go into saturation, except for argon + 5% C_3H_8 and argon + 5% CH_4 in the π/p diagram [LE 81a].

As an example for the use of multiple ionization measurement in an

Fig. 5.27. Mean energy loss, dE/dx, as a function of the particle momentum P, as measured in the drift chamber of the ARGUS experiment [DA 83, SC 84]. The chamber is filled with 96% propane, 3% methylal and 0.7% hydrogen. The relative RMS resolution is $\sigma_r = 4.1\%$. Five kinds of particles can be distinguished in the data.

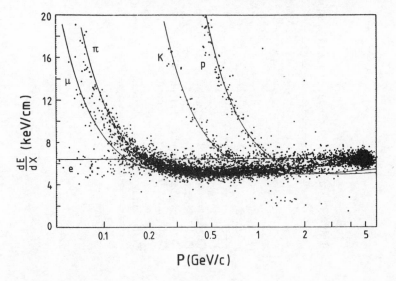

Table 18. *Momentum ranges of particle identifiers*

| Method | Momentum range for π/K separation | | Requirements |
	Fixed target geometry $L=30$ m	Storage ring geometry $L=3$ m	
Time of flight	$P< \quad 4$ GeV/c	$P< 1$ GeV/c	$\sigma_t = 300$ ps
Cherenkov threshold counter	$P< \quad 80$ GeV/c	$P< 25$ GeV/c	10 photoelectrons
DISC Cherenkov	$P< 2000$ GeV/c	—	achromatic gas counter
Ring imaging Cherenkov counter (RICH)	—	$P< 65$ GeV/c	
Multiple ionization measurement	$1.2<P<100$ GeV/c or $P<0.8$ GeV/c	$1.5<P<45$ GeV/c or $P<0.7$ GeV/c	$\sigma_r = 2\%$ for $L=30$ m, $\sigma_r = 3\%$ for $L=3$ m
Transition radiation	$\gamma > 1000$	$\gamma > 1000$	detection of X-rays with $E>10$ keV

Fig. 5.28. Length of detectors needed for separation of π and K mesons.

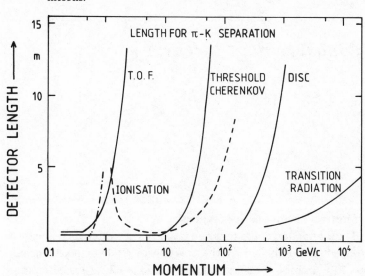

actual experiment, fig. 5.27 shows a two-dimensional distribution of the mean energy loss dE/dx vs. the momentum P of tracks coming from an $e^{+}e^{-}$ reaction at a centre-of-mass energy around 10 GeV. The data are taken by the central drift chamber of the ARGUS detector at the storage ring DORIS II. At momenta below 1.4 GeV/c, the separation of protons, K and π mesons and electrons is possible in this detector.

5.6 Comparison of identification methods for charged particles

The methods discussed in sections 5.2 through 5.5 are useful in certain limited domains of the particle momentum P. If we consider as an example the separation of π and K mesons, below $P = 1$ GeV/c the measurement of ionization energy loss and time-of-flight can be employed; up to $P = 25$ GeV/c threshold Cherenkov counters; from 1.5 to 45 GeV/c measurement of the relativistic rise of energy loss; up to $P = 65$ GeV/c RICH counters; and above $P = 140$ GeV/c transition radiation detectors. The minimum lengths of such detectors needed for π–K separation are estimated in fig. 5.28. Alternatively, the momentum ranges in which π–K separation is possible with a detector of 30 m length (for fixed target experiments) or 3 m length (for storage ring detectors) are shown in table 18.

6

....

Measurement of energy

6.1 Electron–photon shower counters

The interaction of photons and electrons in matter at energies well above 10 MeV is dominated by the creation of electron–positron pairs (section 1.2.2) and by Bremsstrahlung (section 1.2.3). An alternating sequence of interactions of these two types leads to a cascade or 'shower' of electrons, positrons and photons which stops only if the energy of these electrons and positrons approaches the 'critical energy' E_c (section 1.2.3). A simplified picture of this showering process is the following: a primary photon of energy E_0 produces, with 54% probability, in a layer of one radiation length X_0, an e^+e^- pair; these charged particles on average have the energy $E_0/2$. If $E_0/2 > E_c$, these electrons and positrons lose their energy predominantly by Bremsstrahlung. In a layer of thickness X_0, the charged particle energy decreases to $E_0/2e$. On average, a Brems-photon of energy between $E_0/2e$ and $E_0/2$ is radiated. The mean number of particles behind a layer of thickness $2X_0$ is about four. The radiated photons again produce e^+e^- pairs, such that after n generations – corresponding to a thickness nX^0 – there are 2^n particles with average energy $E_0/2^n$, forming a shower. This cascade process stops when the energy loss of electrons by ionization equals the one by Bremsstrahlung, i.e. if the electron energy approaches the critical energy, $E_0/2^n = E_c$. The number of generations up to this point is $n = \ln(E_0/E_c)/\ln 2$, and the number of particles at the shower maximum $N_p = 2^n = E_0/E_c$. The integrated path length S of all electrons and positrons in a shower is approximately

$$S = \frac{2}{3} X_0 \sum_{v=1}^{n} 2^v + s_0 \frac{2}{3} N_p = \left(\frac{4}{3} X_0 + \frac{2}{3} s_0 \right) \frac{E_0}{E_c}. \tag{6.1}$$

Here s_0 is the range of electrons with critical energy. The path length S is nearly proportional to the primary energy E_0.

The path length S is proportional to the total energy E_0 if electrons and positrons can be detected until they come to rest. In practical detectors there is a minimum kinetic energy required for detection (cut-off energy E_k). This effect has the consequence [RO 52] that the visible path length becomes [AM 81]

$$S' = F(z) \frac{X_0 E_0}{E_c} \tag{6.2}$$

with $F(z) \approx e^z [1 + z \ln(z/1.526)]$ and $z = 4.58 \, Z E_k / A E_c$.

Including the effect of the cut-off energy into Monte Carlo calculations [CR 62, NA 65, LO 75] gives the following properties of electron–photon showers:

(i) The number of particles at maximum N_p is proportional to the primary energy E_0.

(ii) The total track length of electrons and positrons S is proportional to E_0.

(iii) The depth at which the maximum occurs, X_{max}, increases logarithmically: $X_{max}/X_0 = \ln(E_0/E_c) - t$, where $t = 1.1$ for electrons and $t = 0.3$ for photons.

The longitudinal energy deposition in an electromagnetic shower can be seen in fig. 6.1 as measured [BA 70] for 6 GeV electrons. A useful

Fig. 6.1. Longitudinal distribution of energy deposition per unit of radiation length, $dE/d\xi$, in an electromagnetic shower, normalized to the energy $E_0 = 6$ GeV of the incident electron. $\xi = x/X_0$ is the depth of the detector in units of radiation length X_0. Measurements (curve) and Monte Carlo calculation (histogram) (after [BA 70]).

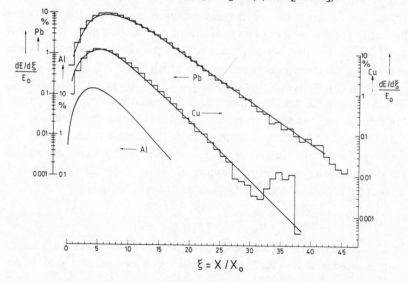

parametrization for this distribution is given by [LO 75]

$$\frac{dE}{dt} = E_0 ct^\alpha \exp(-\beta t) \tag{6.3}$$

where $t = X/X_0$ is the longitudinal depth X in units of X_0, and the parameters $\beta \approx 0.5$, $\alpha \approx \beta t_{max}$ and $c = \beta^{\alpha+1}/\Gamma(\alpha+1)$ vary logarithmically with energy. For proton energies around 1 GeV, the distribution can be approximated by $dE/dt = E_0\, 0.06\, t^2 \exp(-t/2)$ for a lead converter.

The transverse dimension of a shower is determined by the multiple scattering of low-energy electrons. It turns out that a useful unit for transverse shower distributions is the Molière unit $R_M = 21 \text{ MeV} \cdot X_0/E_c$. As shown by the measurements [BA 70] in fig. 6.2, the distribution of shower energy in transverse (radial) bins scaled in R_M is independent of the material used, and 99 % of the energy is inside a radius of $3R_M$.

The energy resolution of an idealized homogenous detector of infinite dimensions is limited only by statistical fluctuations. For a cut-off energy of 0.5 MeV and a critical energy of 11.8 MeV a total track length of 176 cm/ GeV and a resolution $\sigma(E)/E = 0.7\%/\sqrt{(E(\text{GeV}))}$ have been computed [LO 75].

If the shower is not contained in the detector, the fluctuation of the energy leaking out contributes to the resolution. As shown in [DI 80], longitudinal losses induce a larger degradation of the resolution than lateral ones. An estimate for this fluctuation due to longitudinal leakage is $\sigma(E) = (dE/dt)_{t_r} \sigma(t_{max})$, where t_r is the length of the detector and $\sigma(t_{max})$ the fluctuation of the position of the shower maximum. For photons of 1 GeV energy, $\sigma(t_{max}) \sim 1$ and $(\sigma(E)/E)_{leak} = 0.06\, t_r^2 \exp(-t_r/2)$.

Another contribution to the resolution comes from the statistical fluctuation of the number of photoelectrons N_p observed in the detector. If $\alpha_p = N_p/E_0$ is the number of photoelectrons per unit of primary particle energy E_0, this contribution amounts to $(\sigma(E)/E)_{PE} = 1/\sqrt{(\alpha_p E_0)}$.

To these two sources of fluctuations, valid for homogenous calorimeters, we have to add the sampling fluctuations if the shower calorimeter consists of a series of inactive absorber layers of thickness d interspersed with active detector layers ('sampling calorimeter'). If the detectors count only the number of particle traversals, N, the statistical fluctuation in N determines the contribution to the energy resolution. Since N depends on the total track length, $N = S/d = E_0 X_0 F(z)/E_c d$, we obtain [AM 81]

$$(\sigma(E)/E)_{sampl} = \frac{1}{\sqrt{N}} = 3.2\%\, \frac{550}{ZF(z)} \frac{d/X_0}{E_0(\text{GeV})} \tag{6.4}$$

In high Z materials, the lateral dimension of the showers is much larger than in those with low Z, since the Molière unit in units of X_0, $R_M/X_0 =$

21 MeV/E_c, is larger for heavy materials. Consequently the angle θ of electrons and positrons relative to the shower axis is larger [AM 81]. Those shower particles see a larger sampling thickness $d/\cos\theta$, and therefore a smaller number of traversals occur, reducing further the energy resolution by a factor $1/\sqrt{(\langle\cos\theta\rangle)}$. A Monte Carlo calculation [FI 78a] shows that the average $\langle\cos\theta\rangle = \cos(21\text{ MeV}/E_c\cdot\pi)) = 0.57$ for lead. From this calculation, the sampling fluctuation gives $\sigma(E)/E = 4.6\,\%/\sqrt{(E(\text{GeV}))}$ for 1 mm lead sampling thickness and $E_k = 0$.

Another large source of fluctuations enters if the sensitive layers of the calorimeter consist of a gas or a very thin layer of liquid argon ($\lesssim 2$ mm), used as proportional counters. Then low-energy electrons moving at large

Fig. 6.2. Transverse distribution of energy deposition $dE/d\alpha$ in cylindrical intervals around the shower axis, normalized to the energy $E_0 = 6$ GeV of incident electrons; $\alpha = R/R_M$ is the dimension-free ratio of radial distance R from the axis and Molière length R_M. Measurements (dots) and calculation after [BA 70].

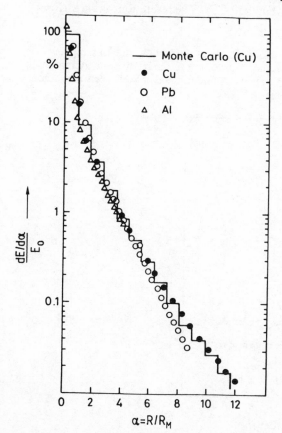

angles relative to the shower axis induce large pulse height fluctuations ('path length fluctuations'), and the Landau tail of the energy loss distribution also leads to a reduction of resolution. The computed effect of these two contributions on the energy resolution of a lead–argon calorimeter can be seen in fig. 6.3. The overall resolution is $18\%/\sqrt{(E(\mathrm{GeV}))}$, more than twice the sampling fluctuation of $7\%/\sqrt{(E(\mathrm{GeV}))}$.

Homogenous shower counters

The best resolutions are obtained with inorganic scintillating crystals. NaI(Tl) detectors with a diameter of $3R_\mathrm{M} = 13\,\mathrm{cm}$ and $15\,X_0 = 40\,\mathrm{cm}$ length have yielded [PA 80] a resolution of $\sigma(E)/E = 2.8\%(E(\mathrm{GeV}))^{-0.25}$ in a large-scale application. For one $24\,X_0$ long counter $\sigma(E)/E = 0.9\%$ $(E(\mathrm{GeV})^{-0.25}$ has been achieved [HU 72]. The new type of crystal, BGO $(B_4Ge_3O_{12})$ gives 8% of the light output of NaI, and a resolution of $\sigma(E)/E = 2.5\%/\sqrt{(E(\mathrm{GeV}))}$ [KO 81]. In more recent tests [BA 85] a matrix of 25 parallelepiped crystals, of size $3 \times 3 \times 24\,\mathrm{cm}^3$, was exposed to electron beams of energies between 1 and 50 GeV. The resolution was $\sigma(E)/E = 4\%$

Fig. 6.3. Calculated contributions of fluctuations due to sampling, track length and Landau distribution to the relative energy resolution of an electromagnetic shower counter made of lead sheets and argon proportional counters [FI 78a].

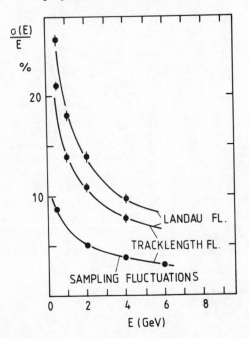

at 1 GeV and $\sigma(E)/E \leqslant 1\%$ at energies between 4 and 50 GeV. The light is measured here by silicon photodiodes.

Lead glass counters detect the Cherenkov light of shower electrons, the resolution is limited by photoelectron statistics. A computation [PR 80], based on 1000 photoelectrons per GeV, gives $\sigma(E)/E = 0.006 + 0.03/\sqrt{(\xi E)}$, ξ being the ratio of photocathode area and counter exit area. Actual measurements [BI 81] with 208 blocks of $36 \times 36 \times 420$ mm^3 give a resolution of $\sigma(E)/E = 0.012 + 0.053/\sqrt{(E(\text{GeV}))}$ for $\xi = 0.35$, in agreement with the calculation.

Sampling shower detectors
The resolution of a lead–scintillator sandwich with 1 mm lead and 5 mm scintillator thickness for a total length of 12.5 radiation length is shown against incident energy in fig. 6.4 [HO 79]. The values for $A = \sigma(E)/\sqrt{E}$ vary from $7\%\sqrt{(\text{GeV})}$ at 100 MeV to $9\%\sqrt{(\text{GeV})}$ at 5 GeV, in agreement with a calculated $5\%\sqrt{(\text{GeV})}$ from sampling fluctuations, $3\text{–}4\%\sqrt{(\text{GeV})}$ from photoelectron statistics, and $2\text{–}5\% \sqrt{(\text{GeV})}$ from leakage.

In lead–liquid argon calorimeters, the ionization is sampled in a proportional mode by the argon chambers defined by two lead plates as electrodes. Resolutions for 2 mm lead plates and 3 mm liquid argon are $\sigma(E)/E = 12\%/\sqrt{(E(\text{GeV}))}$ [KA 81].

A summary on the energy resolution and other figures of merit obtained with electron–photon shower counters is given in table 19.

Fig. 6.4. The quantity $\sigma/\sqrt{(E(\text{GeV}))}$ in units of $(\%\sqrt{(\text{GeV})})$ as a function of the electron energy E for a sandwich shower counter with 1 mm lead plates and 5 mm scintillator sheets. The full line shows the contribution of shower leakage to the resolution [HO 79].

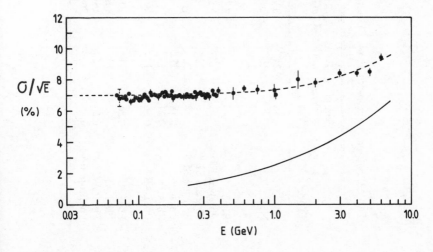

Table 19. *Electromagnetic shower counters*

Type	Sampling thickness X_0	Total thickness X_0	$\sigma(E)/\sqrt{E}$ $\%/\sqrt{(\text{GeV})}$	Spatial resolution (mm)	Angular resolution	Transverse cell size (mm)	Collaboration	Ref.
NaI	—	24	$0.9\,E^{\frac{1}{4}}$				crystal ball	[HU 72]
NaI	—	16	$2.8\,E^{\frac{1}{4}}$					[PA 80]
								[KI 79]
								[CH 78b]
								[KO 81]
BGO		17	2.5					
Pb glass F8	—	12.5	$5.3 + 1.2\,E$	1.3		36×36	IHEP	[BI 81]
Pb glass SF5	—	20	$\sqrt{(6^2 + 2.5^2 E)}$	6	10 mrad	80×104	JADE	[DR 80]
Pb glass SF5	—		$\sqrt{(6^2 + 0.5^2 E)}$	2			NA 1	[NA 1]
Pb-scint.	0.18	12.5	7–9	$11/\sqrt{(E(\text{GeV}))}$		100×100	ARGUS	[HO 79]
Pb-scint.	0.21	13	9	$25/\sqrt{(E(\text{GeV}))}$		200×250	LAPP–LAL	[SC 82]
Pb-LAR	0.36	13.5	10–12	5	5 mrad	$70 \times 70 +$ strips 20 mm	TASSO	[KA 81]
Pb-LAR	0.26	21	10	4	4 mrad	23×23	CELLO	[BE 81]
Pb-LAR		14	11.5				Mark II	[DA 79]
Pb-PWC	0.5	12	16				Mark III	[HI 81]
Pb-prop. tube	1		26	<1		pitch 7.8	NA 24	[BA 84]

Position resolution

The impact point of an electron or photon on an array of shower counters can be obtained by measuring the lateral distribution of energy in the shower. The precision of the position information increases with the number of cells hit by shower particles, and decreases with the cell size. In particular, the accuracy is best if the shower energy is shared equally between two adjacent cells. Binon *et al.* [BI 81], using cells of $36 \times 36 \times 420$ mm^3, have obtained a position resolution of $\sigma_x = 1.3$ mm for 25 GeV electrons. For a lateral cell size $d > 30$ mm, an experimental increase of σ_x is calculated (fig. 6.5) by these authors. On the other hand, a variation of $\sigma_x \propto 1/\sqrt{E}$ has been found [AK 77], confirming the assumption that the spatial resolution depends mainly on the number of shower particles.

With lead–scintillator sandwiches of 10×10 cm^2 lateral dimensions [HO 79], the measured spatial resolution was $\sigma_x = 11$ mm$/\sqrt{(E(\mathrm{GeV}))}$.

6.2 Hadron calorimeters

If a strongly interacting particle ('hadron') of energy above 5 GeV impinges on a block of matter, inelastic interactions, as well as elastic scattering, between the particle and the nucleons in the material take place. In such a collision, several secondary hadrons are produced, e.g. π and K mesons, protons and neutrons. The energy of the primary particle is

Fig. 6.5. Standard deviation $\sigma(y)$ of position resolution for a matrix of lead glass shower counters as a function of the transverse size of blocks. Circle: measurements; full line: calculated resolution assuming homogenous distribution of impact points over matrix; dashed line: impact points at centre of blocks [BI 81].

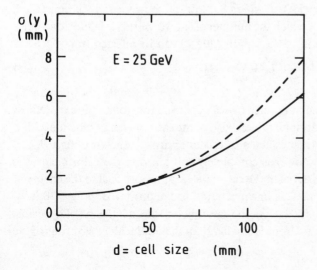

partially transferred to these secondary hadrons, and they in turn can undergo inelastic collisions. Tertiary hadrons are formed, and this cascade process only stops when the hadron energies are so small that they are stopped by ionization energy loss or absorbed in a nuclear process. Such a cascade is called a 'hadronic shower'.

The scale for the spatial development of a hadronic shower, the inelastic production of secondary hadrons, is given by the nuclear absorption length λ. From the inelastic cross-section σ, $\lambda = A/\sigma_i N_0 \rho$ can be obtained. Here A is the mass of one mole of material, σ_i the cross-section for inelastic scattering, N_0 Avogadro's number and ρ the density of the material. Experimental values of λ for materials suitable for calorimetry of hadronic showers are 34 cm (C), 17.1 cm (Fe), 18.5 cm (Pb) and 12.0 cm (U). Compared with the small values for the radiation length of high Z materials enabling the construction of correspondingly small shower counters, the size of hadronic showers is large; typical values for iron calorimeters are 2 m depth and 0.6 m transverse size. The need for such sizes is demonstrated by the measurements shown in figs 6.6 and 6.7 [HO 78b and BL 82]. Fig. 6.6 shows the longitudinal distribution of visible energy of a hadronic shower in an iron–scintillator sandwich with iron plates of 2.5 cm thickness, interspaced with plastic scintillators. The light from five consecutive scintillators is viewed by one photomultiplier (fig. 6.8). The two measurements displayed in fig. 6.6 with incident π mesons of energies 15 GeV and 300 GeV, respectively, indicate how the length of the shower increases with increasing hadron energy. In fig. 6.7, three different measures of shower length are given as a function of the primary hadron energy:

(a) the longitudinal centre of gravity of the shower energy,

(b) the length $L(95\%)$, in which 95% of the shower energy is contained,

(c) the length, for which the number of shower particles falls below unity.

A parametrization of the length $L(95\%)$ can be inferred from fig. 6.7:

$$L(95\%) = [9.4 \ln E(\text{GeV}) + 39] \quad \text{cm Fe} \tag{6.5}$$

In a similar way, the lateral extension of a hadronic shower can be characterized by the transverse size of a calorimeter around the axis defined by the incident hadron for which 95% of the shower energy remains within the calorimeter. Measurements of this quantity are shown in fig. 6.9.

If the hadronic shower is completely contained in the calorimeter, the 'visible' energy registered in the calorimeter is proportional to the energy of the primary hadron. This linear relation is demonstrated in fig. 6.10 from measurements obtained using the calorimeter shown in fig. 6.8. The visible energy for incident hadrons is smaller than that for incident electrons of the same energy.

Apart from the hadron shower particles leaking out longitudinally or laterally, the energy seen in a sampling calorimeter for hadrons is incomplete for several reasons:

(i) there are particles escaping the calorimeter carrying away energy, like muons and neutrinos from pion decay (1% at 140 GeV primary energy),

(ii) there is nuclear excitation and breakup resulting in low-energy γ-rays, neutrons or heavy fragments, which do not reach the sensitive part of the sandwich (20–30% of total energy at 10 GeV for iron calorimeters).

This loss of visible energy, typically 20%, can be seen comparing the light collected from electron- and hadron-induced showers in iron (fig. 6.11) for primary hadrons of an energy between 4 and 300 GeV.

This difference in visible energy ('response') for incident electrons and π

Fig. 6.6. Longitudinal distribution of energy deposited in a hadron calorimeter; Q_p is the energy deposited in counter p consisting of five layers with 2.5 cm iron and 0.5 cm scintillator each; $Q = \sum Q_p$. Measurements for π mesons of 15 GeV and 300 GeV [BL 82].

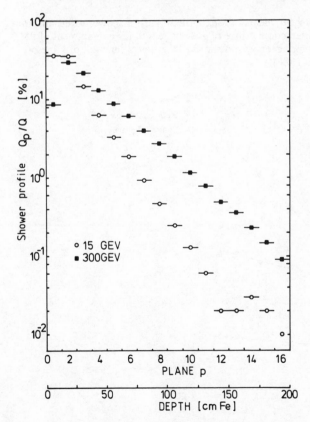

mesons leads to a fluctuation in the visible energy for hadronic showers, because in such a shower the fraction of π^0 mesons fluctuates. Since π^0 mesons decay, with a mean lifetime of 10^{-16} s, to two γ-rays, this component of a hadronic shower develops like an electromagnetic shower. The contribution of such a π^0 meson to the visible energy is larger than that of a corresponding π^+ or π^- meson, and the statistical fluctuation of the number of π^0 mesons therefore leads to the fluctuation of visible energy mentioned above.

On top of this fluctuation there is the sampling fluctuation which alone gives rise to a resolution about twice as large as in electromagnetic showers (see section 6.1). However, the effects of the fluctuation in energy leakage and in the electromagnetic component of the hadronic shower are much larger here and lead to energy resolutions of about

$$\frac{\sigma(E)}{E} \sim \frac{0.9 - 0.7}{\sqrt{(E(\text{GeV}))}} \tag{6.6}$$

if the thickness of material between the sampling devices ('sampling thickness') is below 5 cm of iron.

Fig. 6.7. Measurements of longitudinal hadron shower extension from point of interaction: centre of gravity, length for containment of 95% of the primary energy, and shower length as a function of π meson energy [HO 78b].

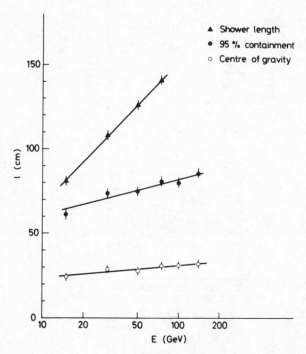

Two ways of improving this resolution have been invented and tried out successfully:

(i) The loss of visible energy through the nuclear excitation and breakup mechanism can be nearly completely compensated by the energy release in nuclear fission of ^{238}U. Energetic neutrons from the fission contribute to the observed signal such that the pulse height for hadron showers becomes nearly equal to the one for electromagnetic showers, as shown [FA 77] in fig. 6.12. The corresponding fluctuations disappear, and the energy resolution decreases by about a factor of two. Experimental results for uranium calorimeters are shown in fig. 6.13; they correspond to

$$\frac{\sigma(E)}{E} = \frac{0.3}{\sqrt{(E(\text{GeV}))}} \qquad (6.7)$$

which is only 50% higher than the lower limit given by sampling fluctuations.

(ii) Another method [DI 79, AB 81] reduces the fluctuation due to the electromagnetic component by weighting the response in individual counters. Electromagnetic parts of the shower are localized, therefore producing very large depositions in individual counters. If the measured

Fig. 6.8. Schematic layout of an iron–scintillator calorimeter [BL 82].

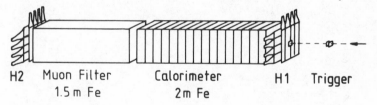

H2   Muon Filter   Calorimeter   H1   Trigger
         1.5 m Fe   2 m Fe

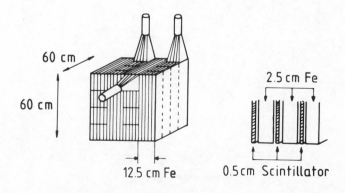

60 cm

60 cm

12.5 cm Fe

2.5 cm Fe

0.5 cm Scintillator

response in one counter E_k is corrected downwards for large response, $E'_k = E_k(1 - CE_k)$, then the resulting resolution in the sum $\sum E'_k$ is markedly improved over the one in $\sum E_k$, as shown [AB 81] in fig. 6.14 for 2.5 cm Fe sampling calorimeter exposed to π mesons of 100, 200 and 300 GeV energy. The resolution can be approximately described by

$$\frac{\sigma(E)}{E} = \frac{0.58}{\sqrt{(E(\text{GeV}))}} \tag{6.8}$$

and the measured values for $\sigma(E)/\sqrt{E}$ as a function of E are given in fig. 6.15.

The energy resolution of hadron calorimeters varies with the thickness d ('sampling thickness') of absorber plates. Since the number N of ionizing particles registered in the counters between the plates is inversely

Fig. 6.9. Radial extension of hadron showers for containment of 95% of primary energy as a function of distance from the interaction point [HO 78b].

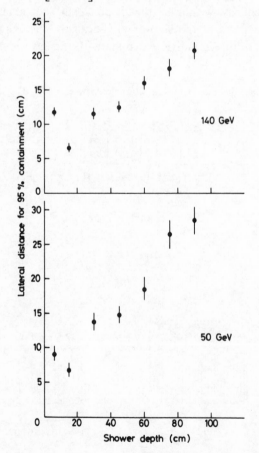

140 GeV

50 GeV

Lateral distance for 95% containment (cm)

Shower depth (cm)

Fig. 6.10. Visible energy measured in the calorimeter shown in fig. 6.8, in units of minimum ionizing particles ('nep') [BL 82].

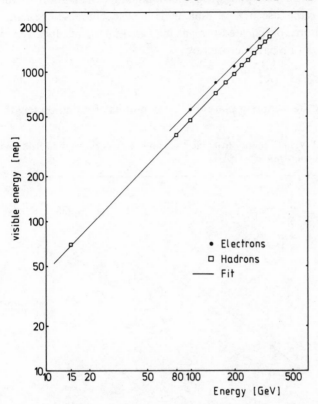

Fig. 6.11. Ratio of visible energies of π mesons and electrons at the same energy ('response ratio') in iron–scintillator calorimeter [BL 82].

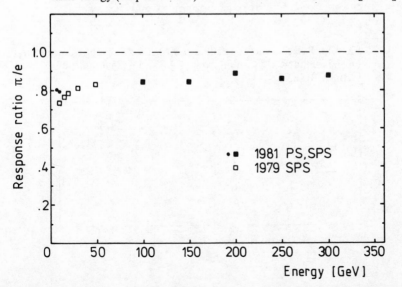

proportional to d, one would expect the relative resolution to vary with $\sqrt{d}$ if sampling fluctuations would yield the dominant contribution to the resolution (as it is the case for electromagnetic shower counters). The actual dependence is more complicated here, as indicated by the results given in fig. 6.16. These can be parametrized by

$$\frac{\sigma(E)^2}{E} = 0.25 + (R')^2 \frac{4t}{3} \tag{6.9}$$

Here $t = d/X_0$ is the sampling plate thickness in units of radiation length,

Fig. 6.12. Visible energy from electron and hadron showers in iron and uranium calorimeters [FA 77].

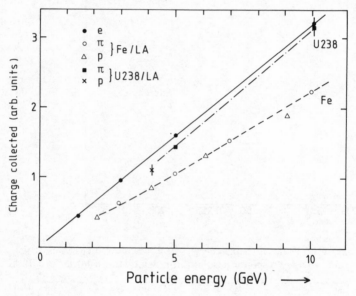

Fig. 6.13. Relative energy resolution $\sigma(E)/E$ for different calorimeters; Fe (1.5 mm) and U (1.7 mm) from [FA 77]; Fe (25 mm) from [AB 81, BL 82].

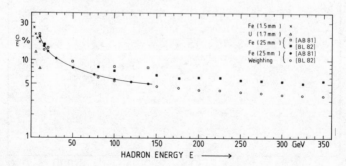

and R' is a free parameter, determined from the data to be $R' \sim 0.3$–0.4. The measured resolutions indicate that a considerable improvement of the resolution cannot be achieved by reducing the plate thickness below $d = 2$ cm of iron, and that the limit for $d \to 0$ is $\sigma(E)/\sqrt{(E(\mathrm{GeV}))} \sim 0.5$.

The sampling of ionization in hadron calorimeters can be done by scintillators, liquid argon ionization chambers, proportional chambers, or flash tubes. The choice between these detectors depends on the desired resolution, granularity and cost. For moderate-sized geometries, liquid argon chambers and scintillators are used for best resolution. For very large fine grain calorimeters (ν–e scattering, proton decay), the proportional tubes or flash tubes give granularities down to 5 mm × 5 mm at a price which still allows the construction of multi-hundred tonne calorimeters.

6.3 Calibration and monitoring of calorimeters

In a typical large-scale calorimeter there will be several thousand channels of analog pulse height information which is converted to digits

Fig. 6.14. Pulse height distribution from iron–scintillator hadron calorimeter for mono-energetic π mesons. Dotted histogram: distributions without weighting; full histograms: weighted distributions [BL 82].

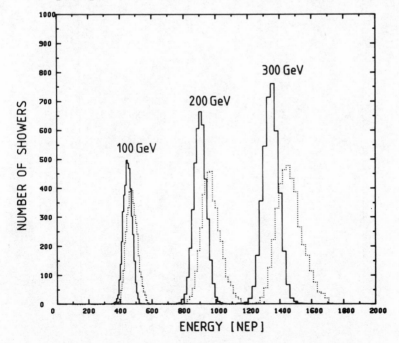

and registered. A severe problem with such a number of channels is their calibration and monitoring.

The calibration can be done by using suitable hadron beams and calibrating the response of the calorimeter, where for each sampling detector the pulse height is measured in terms of minimum ionization deposited by high-energy muons.

If there are not as many muons in each sampling detector as are needed for day-to-day monitoring, another source of calibrated pulse heights is needed. For liquid argon calorimeters, such a source is obtained by depositing a known amount of charge into the ion chamber. The same can be done for proportional chambers.

For calorimeters using scintillation counters, monitoring has to test not only the stability of electronic pulse processing from the output of the

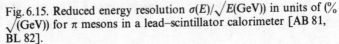

Fig. 6.15. Reduced energy resolution $\sigma(E)/\sqrt{E(\text{GeV})}$ in units of (% $\sqrt{(\text{GeV})}$) for π mesons in a lead–scintillator calorimeter [AB 81, BL 82].

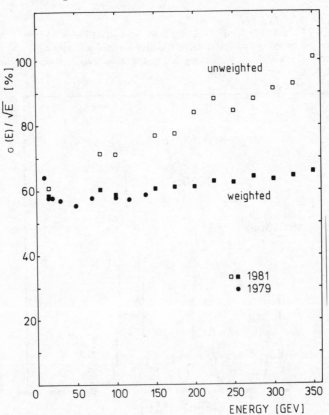

photomultiplier up to the register for the digital pulse height, but also the stability of the photomultiplier gain. For this purpose, a calibrated stable light source is needed. Pulsed nitrogen lasers emitting at a wavelength $\lambda =$ 337 nm have been used in such monitoring systems. The light has to be distributed to a few thousand counters, and the light intensity has to be calibrated against a standard light source.

In one of such systems [GR 80], built for the Underground Area no. 1 (UA1) experiment at the CERN proton–antiproton collider, the laser light is injected into a rectangular box covered inside with a highly reflective material (millipore). After many reflections inside, the light is diffuse and leaves the box through quartz fibres of 200 μm diameter. This fibre has attenuations of $\sim$ 400 dB/km in the ultraviolet. Each of the 8000 fibres is connected to a plexiglas prism glued to the centre of the scintillator. The ultraviolet light pulse then produces scintillation light which reaches the photomultiplier and leads to a digital pulse height.

In another system [EI 80], designed for the neutrino experiment of the CDHS collaboration, the laser beam passes through a filter, is widened up optically, and then illuminates a scintillator piece glued onto a plexiglas rod (fig. 6.17). The blue POPOP light emitted isotropically from the scintillator travels down the rod by internal reflection and is partially accepted by the

Fig. 6.16. Reduced energy resolution $\sigma(E)/\sqrt{E(\text{GeV})}$ in units of (% $\sqrt{(\text{GeV})}$) as a function of sampling thickness of iron plates.

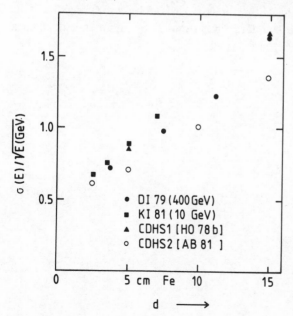

Fig. 6.17. Laser calibration system for 2304 photomultipliers in a hadron calorimeter [EI 80].

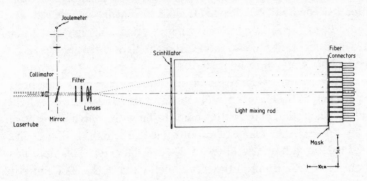

2304 fibres grouped into 144 bundles of 16, each bundle in one connector. The homogeneity of illumination of the fibres is within 1%. A mechanical mask moving across the front of the connectors permits one group of 192 fibres at a time to be illuminated. This is required by the number of ADC channels available. The transmission of the fibres of 200 μm diameter (QSF 200 A) is 180 dB/km for the blue scintillator light, such that over a length of 25 m the attenuation is a factor of 2.8. Each fibre is connected to a light guide through a small (4 mm dia.) cylindrical rod. With this system, by exchanging filters of different density on a 'filter wheel', the linearity of all tubes can be measured in a dynamic range from 1 to 2000 times minimum ionization. The absolute calibration is carried out by comparing one of the fibre outputs with the standard light from an α-source embedded in a scintillator (see fig. 1.16).

7

Measurement of momentum

7.1 Magnet shapes for fixed target experiments

In a fixed target interaction, the reaction products are usually concentrated in a cone around the incident beam direction (z) because of their limited transverse momenta and the Lorentz boost for longitudinal momenta. The opening angle of this cone is given by the ratio of the mean transverse momentum (~ 300 MeV/c) and the longitudinal momentum of the particle. For a secondary particle of high total momentum with the components (P_x, P_y, P_z) we have $P_x, P_y \ll P_z$. If a particle with charge e traverses a homogeneous magnetic field with components $(0, B_y, 0)$ and a length L, its orbit in the field is a circle with radius $R = P/eB_y$. The angular deflection θ in the (x, z)-plane can be inferred from fig. 7.3:

$$2 \sin \frac{\theta}{2} = \frac{L}{R} = -\frac{eB_y L}{P} \qquad (7.1)$$

The change in transverse momentum is

$$\Delta P_x = P \sin \theta \approx -eB_y L = -e \int B_y \, dz \qquad (7.2)$$

in good approximation for small deflection angles. For an inhomogeneous field, the product $B_y L$ has to be replaced by the 'field integral' $\int B_y \, dz$. Numerically, a field integral of $10\,\text{kG}\,\text{m} = 1\,\text{T}\,\text{m}$ gives a transverse momentum change of $\Delta P_x = 0.3$ GeV/c. In this approximation, the knowledge of the angular deflection θ and of the field integral are sufficient for a measurement of the momentum P. The angular deflection in this approximation can be obtained from the relation

$$\sin \theta = \sin(\theta_2 - \theta_1)$$
$$= \sin \theta_2 \cos \theta_1 - \sin \theta_1 \cos \theta_2 \approx \sin \theta_2 - \sin \theta_1$$

where θ_2 is the angle of the outgoing particle in the (x, z)-plane and θ_1 the same for the incoming particle.

If the magnetized volume is evacuated and the multiple scattering in the position detectors is neglected, the error in momentum, $\sigma(P)$, comes from measurement error $\sigma(x)$ in the chambers alone

$$\frac{\sigma(P)}{P} = 2\,\frac{P}{\Delta P_x}\,\frac{\sigma(x)}{h} \tag{7.3}$$

if the lever arm for the angle measurement before and after the magnet is h. For a field integral of $50\,\mathrm{kG\,m}$, $\sigma(x) = 0.3\,\mathrm{mm}$ and $h = 3\,\mathrm{m}$ this gives $\sigma(P)/P \sim 1.3\%$ for $P = 100\,\mathrm{GeV}/c$, or $\sigma(P)/P^2 = 1.3 \times 10^{-4}/(\mathrm{GeV}/c)$.

These 'air core' magnets come in different forms (fig. 7.1): H-magnets have symmetrical flux return yokes, C-magnets asymmetrical ones (and a less uniform field). The amount of iron in the flux return depends on the desired field strength in the air gap. For a cubic magnetized region, the volume of iron needed, V_{Fe}, relative to the magnetized air gap volume V_{Mag} for different field strength B in the gap is approximately given by

$$\frac{V_{\mathrm{Fe}}}{V_{\mathrm{Mag}}} \sim \left(2 + \frac{B}{B_s}\right)\frac{B}{B_s}$$

as shown by the curve ACM in fig. 7.2. If B has to reach the saturation field

Fig. 7.1. Magnet shapes for fixed target experiments: (a) H magnet with air gap; (b) C magnet with air gap; (c) iron core toroid magnet; (d) iron core H magnet.

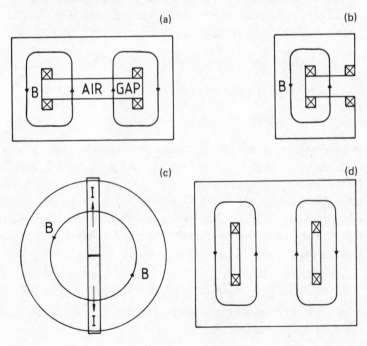

strength B_s, then $V_{Fe}/V_{Mag} \sim 3$, which is very uneconomical. More usual magnets have $B/B_s \sim \frac{1}{2}$ to $\frac{1}{3}$.

If the particles to be analysed are high-energy muons, a more economical form of magnets are 'iron core' magnets (ICM). Here the field lines stay completely within iron, either in the form of a toroid, where the field lines are circular around a central hole for the coils, or in a kind of H-magnet, where the central region is also filled with iron (fig. 7.1). The momentum resolution here is limited by multiple scattering of the muons in iron and the measurement error of the muon track. Multiple scattering results in a mean transverse momentum change of

$$\Delta P_T^{MS} = 21 \text{ MeV}/c \left(\frac{L}{X_0}\right)^{\frac{1}{2}} \tag{7.4}$$

where L is the length of iron traversed. The momentum error induced by multiple scattering in the x-direction is given by

$$\left|\frac{\sigma(P)}{P}\right|^{MS} = -\frac{\Delta P_x^{MS}}{\Delta P_x} = \frac{15 \text{ MeV}/c \sqrt{(L/X_0)}}{e \int B_y \, dz} = 0.26 \frac{1}{\sqrt{L}} \tag{7.5}$$

where L is measured in metres and $B_y = 1.5$ T is assumed. This error is independent of the particle momentum P, and for $L = 5$ m it amounts to $[\sigma(P)/P]^{MS} \sim 12\%$.

Fig. 7.2. Volume of iron V_{Fe} needed per magnetized usable volume V_{mag} for iron core magnets (ICM) and air core magnets (ACM). The ratio V_{Fe}/V_{mag} is plotted vs. magnetic flux density B in units of saturation flux density B_S.

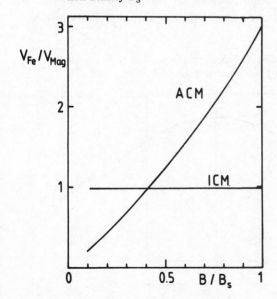

In addition to this error from multiple scattering, there is the contribution from track measurement errors. If the position of a muon in an iron core magnet of length L is measured at three equidistant points along the track, the sagitta of the circular orbit is (see fig. 7.3):

$$s = R - R \cos \frac{\theta}{2} \approx \frac{R\theta^2}{8}$$

In the SI system, B is measured in tesla, the radius R in metres and the momentum in GeV/c. Then $R = P/0.3B$ and $\theta \approx 0.3\, BL/P$. For small deflection angles we have

$$s = 0.3 \frac{BL^2}{8P}$$

Since the sagitta s is determined by three measurements to a precision of $\sigma(s) = \sqrt{(3/2)}\sigma(x)$, we obtain

$$\left| \frac{\sigma(P)}{P} \right|^{M} = \frac{\sigma(s)}{s} = \frac{\sqrt{(3/2)}\sigma(x)8P}{0.3\, BL^2} \qquad (7.6)$$

If the track, instead, is measured at N equidistant points, the corresponding relation is [GL 63]:

$$\left| \frac{\sigma(P)}{P} \right|^{M} = \frac{\sigma(x)P}{0.3\, BL^2} \sqrt{\left(\frac{720}{N+4} \right)} \qquad (7.7)$$

For $B = 1.5$ T, $L = 5$ m, $\sigma(x) = 1.5 \times 10^{-3}$ m and $N = 6$ this gives $[\sigma(P)/P]^{M} \approx 11\%$ at a momentum of 100 GeV/c, about the same magnitude as the contribution from multiple scattering. Fig. 7.4 shows these two contributions to the momentum resolution for iron core magnets of different length. If measurements are taken at equidistant positions along

Fig. 7.3. Bending radius R, deflection angle θ and sagitta s for a charged particle in a homogenous magnetic field of length L.

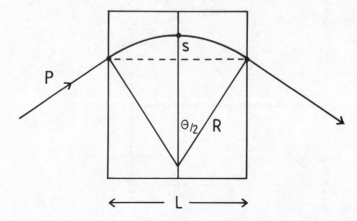

the track, the momentum error from measurement errors decreases with $L^{-\frac{5}{2}}$ because not only the field integral and the lever arm for angle measurements increase with L, but also the number of measurements along the track.

7.2 Magnet shapes for storage ring experiments

For these experiments, the laboratory system is also the centre-of-mass system of the reactions. The interaction rates for the interesting processes are very low. For example, electron–positron storage rings in the GeV energy range reach luminosities around $L = 10^{30}/cm^2$ s, which leads, at a cross-section of the order of nanobars, to rates of $10^{-3}/s$. The rates for

Fig. 7.4. Relative momentum resolution $\sigma(P)/P$ for muons of momentum P_μ measured with an iron core magnet of length L at $B = 1.5$ T. Contributions from measurement error and from multiple scattering are shown separately.

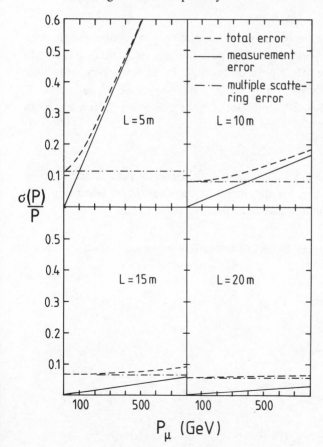

weak processes at antiproton–proton colliders are even smaller. It is therefore necessary to cover the total solid angle around the interaction point by the detector. For the magnetic field in such a detector, the following geometrical arrangements can be realised (fig. 7.5):

(*a*) Dipole magnet in connection with two compensator magnets
The field integrals of the three magnets are adjusted in such a way as to keep the beam particles on orbit. This configuration has the advantage of a homogeneous field around the interaction region. The momentum resolution is best for secondary particles emitted backward or forward relative to the beam axis. The momentum of particles emitted at 90° relative to the beam axis and parallel to the field lines cannot be measured. Since beam particles are also deflected in this scheme, such fields cannot be used in electron–positron machines because of the strong synchrotron radiation produced by the bending.

(*b*) Dipole magnet with split field
Again, the most precise momentum measurement is possible for particles emitted forward or backward. For particles emitted transversely to the beam direction, the orbits proceed through an inhomogeneous field, the reconstruction of tracks is complicated. The transverse field, as in case (*a*), impedes the use of such a configuration for electron–positron machines because of synchrotron radiation.

(*c*) Toroidal field
The inner conducting cylinder causes multiple scattering of tracks coming from the interaction point, deteriorating the momentum resolution. The beams penetrate the detector in a field-free region.

Fig. 7.5. Magnet shapes for storage ring experiments; full lines indicate currents.

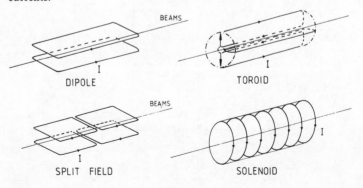

(*d*) Solenoid

The magnetic field lines are parallel to the beam particles, thus avoiding forces on the beam particles and production of synchrotron radiation. The best momentum resolution is obtained for particles emitted perpendicular to the beam axis. It is necessary to close the iron yoke for the return path of the magnetic flux by two end caps on both sides of the cylinder. The part of the detector inside the coil is only accessible after removal of these end caps.

For proton (and antiproton) storage rings, split field and dipole magnets are being used as well as solenoids, while for electron–positron rings the solenoid has been most widely chosen. In general, storage ring detectors resemble each other much more closely than the detectors in fixed target machines.

7.3 Central tracking detectors for storage ring experiments

For the solenoid fields used in storage rings, momentum measurement is usually done in the central tracking detector around the interaction point. This detector has a cylindrical shape with cylinder co-ordinates r (radius), φ (azimuthal angle), and z along the magnetic field which is parallel to the cylinder axis. If the measurement error in the (r, φ)-plane perpendicular to the field is $\sigma_{r\varphi}$, the momentum component in that plane P_T, is measured with an error [GL 63]

$$\left[\frac{\sigma(P_T)}{P_T}\right]^{M} = \frac{\sigma_{r\varphi} P_T}{0.3 \; BL^2} \sqrt{\frac{720}{N+4}} \tag{7.8}$$

where B is the flux density in tesla, L the radial track length in metres and N the number of measured points along the track at uniform spacing.

In addition to this measurement error, there is the error due to multiple scattering

$$\left[\frac{\sigma(P_T)}{P_T}\right]^{MS} = \frac{0.05}{BL} \sqrt{\left(\frac{1.43L}{X_0}\right)} \tag{7.9}$$

For reconstructing the total momentum of the track, the polar angle θ in a plane containing the cylinder axis also has to be measured.

The modulus of the momentum is then

$$P = \frac{P_T}{\sin \theta} \tag{7.10}$$

For the measurement of the polar angle θ, again two sources contribute to the total error, σ_θ. One of them is the measurement error of the z-co-ordinates of N points measured along the track of length L, which causes an

Table 20. *Central track detectors*

Name	Ref.	Max. track length Radial L (cm)	Max. track length Axial z (cm)	Flux density B (T)	No. of measured points	Gas pressure (bar)	No. of signal wires	Spatial resolution σ(r, φ) (μm)	Spatial resolution σ_z (mm)	Method of z measurement	Momentum resolution σ/P² [(GeV/c)⁻¹ %] calc.	Momentum resolution σ/P² [(GeV/c)⁻¹ %] meas.
TASSO	[BO 80a]	85	330	0.5	15	1	2340	200	3–4	4° stereo		1.7
CELLO	[BE 81]	53	220	1.3	12	1	6432	170	0.44	cathodes		5
CLEO	[ST 81b]	75	190	0.5 (1.5)	17	1		250	5 (0.25)			
Mark II	[DA 79]	104		0.4	16	1		200	4			1.9
JADE	[DR 80]	57	234	0.45	48	4	1536	180	16	charge division		2.2
AFS	[CO 81]	60	128	0.5	42	1	3400	200	17	charge division		
UA 1	[BA 80]	112	250	0.7	~100	1	6100	drift: ch. div.	250 μm 8–25 mm	charge division		
TPC	[NY 81]	75	100	1.5	186	10	2232 +13 824 +144	≤200	0.2	drift	1.0	
TRIUMF	[HA 81a]	54	69	0.9	12	1	+630	(600)	(0.6)	drift		

angular error

$$(\sigma_\theta)^M = \frac{\sigma_z}{L}\left(\frac{12(N-1)}{N(N+1)}\right)^{\frac{1}{2}} \qquad (7.11)$$

The other source comes from multiple scattering of the track and is

$$(\sigma_\theta)^{MS} = \frac{0.015}{3^{\frac{1}{2}}P}\left(\frac{L}{X_0}\right)^{\frac{1}{2}} \qquad (7.12)$$

It appears from these relations that the momentum resolution improves with the product BL^2 if the measurement error is the dominant contribution. An increase in the number of points N measured along the track only gives an improvement with $\sqrt{N}$.

The types of drift chambers described in sections 3.3–3.5 have been used as central tracking detectors in storage ring experiments. The properties and figures of merit of some of these detectors operational in 1982 are contained in table 20. Most of these detectors were part of experiments detecting e^+e^- reactions at centre-of-mass energies $\sqrt{s}$ up to 40 GeV, one of them for p–p scattering experiments at $\sqrt{s} \leqslant 60$ GeV, and one for detecting $\bar{p}p$ collisions at $\sqrt{s} = 540$ GeV (UA1). Since a proton consists of three quarks carrying a momentum fraction $x \sim \frac{1}{3}$ of the proton momentum, the mean centre-of-mass energy available for quark–quark scattering is $\sqrt{(x_1 x_2 s)}$, where x_1 and x_2 are the momentum fractions of the constituents. The effective centre-of-mass energy is therefore about 100 GeV for the latter experiment.

For future detectors at the large electron–positron collider LEP and CERN, with centre-of-mass energies up to 200 GeV, central detector magnets with much larger values of the bending power BL^2 are being constructed. As an example, the superconducting solenoid of the Aleph collaboration [AL 83a] has a diameter of 4 m and a flux density of 1.5 T, i.e. $BL^2 = 6$ T m². In comparison with the momentum resolutions quoted in table 20, namely $(\sigma_P/P^2) = (1–5)\%$ (GeV/c)$^{-1}$, such a magnet will enable a resolution of $(\sigma_P/P^2) = 10^{-3}$ (GeV/c)$^{-1}$ if the tracks are measured at 10 points in the $(r\varphi)$-projection with a precision $\sigma_{r\varphi} \sim 150\ \mu$m. With this precision, the sign of a particle with 300 GeV/c momentum can be determined with 90% confidence. If at a later date the LEP tunnel should be used for a proton–antiproton collider with 10 T magnets and a centre-of-mass energy of 9 TeV, even these LEP detectors will not be adapted to the detection of particles and radiation from collisions at this super-high energy. Detector development is therefore necessary and a continuous process.

8

Applications of detector systems

As there is an abundance of applications of radiation detectors, only a few examples can be mentioned here. They are selected from the fields of medicine, space missions, geology to atomic, nuclear and particle physics. In medical applications, radioactive nuclides are used to determine the sizes and functions of inner organs. Active nuclides are incorporated and concentrated in certain regions of the body. Their γ-radiation is measured and localized. In geophysics, natural or induced γ-radiation is used as an indicator for the composition of geological layers when searching for petroleum or minerals. In space aeronautics, the measurement of fluxes of charged particles and γ-rays have immediate importance as a radiation dose monitor for the astronauts themselves. On the other hand, the knowledge of the particle fluxes from the sun or from our galaxy is an important indicator for astrophysical processes. For experiments in atomic and nuclear physics, many of the detectors described above have been developed, e.g. the proportional counter and the solid state detector. High-energy particle physics has used and further developed these methods and invented new ones. Examples for this are the multiwire proportional chamber, the drift chamber and the Cherenkov counter. Since high-energy physics searches for the elementary constituents of matter, experiments have to probe these constituents down to dimensions of 10^{-18} m. The centre-of-mass energy has to be correspondingly high, say 100 GeV. As a consequence, detector systems have to be large (dimensions $\geqslant 10$ m), massive (mass $\lesssim 2000$ t) and complex (10^5 channels of analogous information). The highest centre-of-mass energies $\sqrt{s}$ are obtained at the moment in $\bar{p}p$ collisions at CERN ($\sqrt{s} \leqslant 620$ GeV). In e^+e^- reactions, the upper limit reached at the PETRA storage ring is $\sqrt{s} = 42$ GeV. At the LEP e^+e^- collider, $\sqrt{s} = 100$ GeV is planned for 1989, and 200 GeV later. The highest proton energy in the laboratory system is achieved at the superconducting tevatron at the Fermi Laboratory near Chicago. This technique of superconducting coils for dipole and quadrupole magnets will also be used for the electron–proton colliding machine HERA at DESY in Hamburg.

8.1 Medical applications

A well-known use of radionuclides is based on the concentration of iodine in the thyroid gland. If a patent medicine containing ^{125}I is dispensed to a patient, the evolution of the metabolism in, and the size of, the thyroid gland can be measured with the help of the γ-radiation emitted by the nuclide. The detector is usually a NaI(Tl) crystal registering the intensity of γ-radiation at different positions relative to the thyroid gland. If the intensity is measured at many equidistant points in a rectangular co-ordinate system, a 'scan' results; and the instrument is called a 'scanner'.

Other inner organs of the body can be imaged by γ-ray detectors if a drug which is preferentially deposited in that organ during metabolism contains the radionuclide technetium 99 metastable (^{99m}Tc). With the help of a γ-ray scanner, tumours in the brain, in the liver and in the bones can be localized. The time needed for a scan can be reduced considerably if the γ-radiation is registered simultaneously by not just one but by hundreds of scintillators each with one photomultiplier. Alternatively, one large scintillator can be used which is viewed by a two-dimensional matrix of photomultipliers. The direction from which γ-radiation can impinge on the scintillator is selected by a suitable arrangement of collimators with parallel or converging channels. Two scanners, one above and one on top of the reposing patient, each give independent information about the distribution of the nuclide. The spatial image of this distribution is then measured in a few minutes and can be displayed with the help of a computer. Spatial resolutions of 5 mm can be achieved. This method is particularly advantageous if the time development of a physiological process, measured in this way, can give information about the illness, e.g. for heart diseases.

Three-dimensional images of an inner organ can be obtained with scanners if the γ-ray detectors are arranged in such a way that they measure the γ-radiation from the organ in three directions. An example of such an instrument, used for brain tomography, is shown in fig. 8.1 [KU 76]. This detector consists of 4×8 NaI(Tl) crystals of $2.5 \times 7.6 \times 2.5 \text{ cm}^3$ volume, with collimators aligned towards the centre of the symmetric setup. The four detector arms rotate with a period of 50 s around the symmetry axis of the instrument, where the examined brain is situated. If the signals of all 32 counters are measured during five revolutions of the scanner, the computer connected to the instrument composes the picture of the spatial distributions within five minutes. Such a display is shown in fig. 8.2. It shows the concentration of ^{99m}Tc in a brain tumour with a resolution of 1.7 cm.

A further improvement of medical diagnostics is possible by the detection of *two* γ-rays of 511 keV energy emitted during the annihilation of positrons with electrons of the body. Suitable positron-emitting radionuclides are ^{11}C, ^{13}N, ^{15}O and ^{86}Rb. The advantage of this method comes from the property of annihilation that the two γ-quanta are emitted at 180° relative to each other. If both quanta are registered a straight line is defined between the impact points on which the point of annihilation lies. Collimators in front of the γ-ray detectors are not needed here. Investigations of the heart, the lungs and of the brain are possible with this method. For example, the nuclide ^{11}C can be incorporated by the patient by inhaling ^{11}CO, which then enters the blood circulation as ^{11}C-carboxyhaemoglobin.

It is possible to replace the NaI crystals used as γ-ray detectors by the new material BGO which has a superior energy resolution (see sections 6.1 and

Fig. 8.1. Brain tomograph Mark IV; each of the four detector arms A, B, C, and D contains eight NaI(Tl) crystals [KU 76].

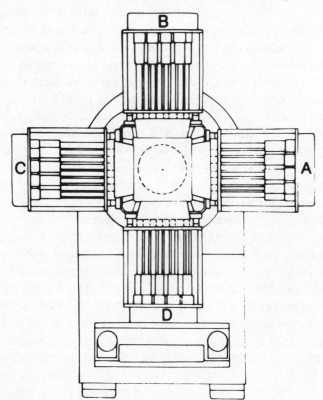

4.2). A review of tomographic methods is given by Phelps [PH 77] and by Heath *et al.* [He 79].

8.2 Geophysical applications

Radiation detectors are used mainly in geographical applications for two purposes, namely searching for uranium-rich minerals at the earth surface and seeking sediment layers containing petroleum or uranium deep underground.

The first method involves searching for the natural radionuclides ^{40}K, ^{232}Th and ^{238}U at the surface from a low-flying aircraft. These nuclides are traced by their monoenergetic γ-radiation at energies of 1.46 MeV, 2.62 MeV and 1.76 MeV, respectively. For this purpose large γ-counters, with volumes of NaI up to 50 l, and multichannel pulse height analysers are installed on the plane. The measured intensities of the characteristic γ-lines allow a determination of uranium concentrations in the area investigated [GR 75].

Fig. 8.2. Concentration of ^{99m}Tc in a brain tumour measured with the tomograph in fig. 8.1 [KU 76].

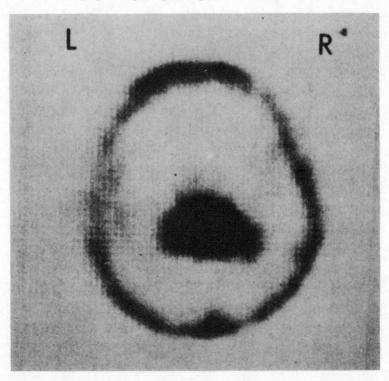

With the second method, the geological structure of the materials surrounding a bore hole is explored as a function of depth by lowering a γ-ray detector into the hole. The detector then records the natural γ-activity of the nuclides ^{40}K, ^{232}Th and ^{238}U contained in the rock formations perforated by the drill, which allows conclusions about the presence of certain minerals. The effectivity of the method is improved by combining the counter with an artificial neutron or γ-ray source, from which the counter is shielded such that the radiation cannot reach the counter directly. Fig. 8.3 shows a sketch of such a well-logging tool. The γ-rays or neutrons from the source will, amongst other processes, be scattered from the material surrounding the hole or they will induce nuclear fluorescence. The γ-rays emitted by the excited nuclei in the rock material can then be detected by the γ-counter. Alternatively, the probability of scattering depends on the density and the atomic number of the scattering material, and a depth profile of this probability for one or more γ-ray sources yields

Fig. 8.3. Probe for well-logging using γ-ray source (principle).

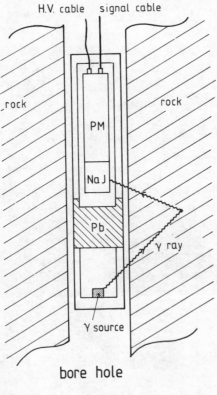

bore hole

ϕ 8 cm

important information on concentrations and spatial contributions of certain elements in these layers.

As a neutron source for such devices, one uses either the combination of an α-source with beryllium or even a small deuteron accelerator, which, through a (d, t)-reaction, liberates neutrons with energies in the MeV range. The advantage of the latter method lies in the pulsed mode of operation of the accelerator, enabling the delay time between neutron emission and γ-fluorescence radiation to be measured. In this way, γ-rays from inelastic neutron scattering can be distinguished from γ-rays due to neutron capture reactions [HE 78]. This procedure is highly selective for neutron-induced reactions in particular elements, and is therefore used to measure ratios of concentration of C vs. O and Si vs. Ca. The method then allows to distinguish between oil and water in porous beds.

8.3 Applications in space sciences

One of the first discoveries made with particle detectors in space missions was the observation of a large density of charged particles at a relatively small distance from the earth during the 1958 flight of the spacecraft Explorer I. With one single Geiger–Mueller counter it was found that the earth is surrounded by two radiation belts, at distances of 5000 km and 20 000 km from the earth's surface. The inner of these 'van Allen-belts' consists mainly of protons, the outer one of electrons. The belts are built up by particles from the solar wind captured by the earth's magnetic field. The belts do not constitute a danger to astronauts since they are crossed by the spacecraft in a short time. Also, the geostationary satellites are not affected by these belts since they are stationed 36 000 km from the centre of the earth, well outside the outer belt.

The intensity of cosmic particle radiation and of the electrons and protons from the sun constituting the solar wind are measured from high-flying balloons, from satellites, or, in recent years, from the reusable Spacelab. Also, cosmic γ-ray and X-ray radiation have been observed from satellites. The first observation of such cosmic γ-radiation [ME 64] was based on measurements with a CsI(Tl) crystal installed on the spacecraft Ranger. At a distance of 10^5 km from earth, background radiation of solar origin did not disturb the measurement. Subsequent experiments in which the direction of the γ-rays was measured have shown that our galaxy contains many sources of such radiation. In some of these sources the intensity of radiation pulsates periodically. The investigation of these γ-ray sources opened up a new field of astronomy. In particular, a dedicated satellite with a large γ-ray telescope was due for launching into orbit in 1985 [FI 78a]. It carries a NaI(Tl) crystal with a volume of $76 \times 50 \times 20$ cm^3 and an arrangement of spark chambers with converters between them. An incident γ-ray is converted to an e^+e^- pair which is detected in the spark chambers. After this measurement of the direction of incidence of the γ-ray, the e^+e^- pair impinges on the NaI crystal, and the electromagnetic shower energy is measured. The instrument is designed for γ-ray energies between 1 and 20 MeV; in the energy range from 0.02 to 1 MeV only the energy is measured.

A further application of radiation detectors in space science is the investigation of the chemical composition of the moon's surface. This has been achieved by measurement of the energy spectrum of γ-rays from the moon from the spacecraft Apollo 16 while on orbit around the moon. A NaI crystal (7 cm dia., thickness 7 cm), viewed by a photomultiplier with 75 mm cathode diameter, was surrounded by a plastic scintillator serving as an anticoincidence shield against charged particles (fig. 8.4). In order to obtain the energy spectrum of γ-rays originating from the surface of the moon, background from other sources has to be subtracted. This background comes from interactions of charged particle cosmic rays with material in the spacecraft, from Bremsstrahlung of cosmic and solar electrons and similar origins. After subtraction of this background, the energy spectrum shown in fig. 8.5 is obtained; it extends from zero to an energy of about 8 MeV. Apart

Fig. 8.4. NaI(Tl) γ-ray spectrometer on board Apollo 16 spacecraft [HA 74].

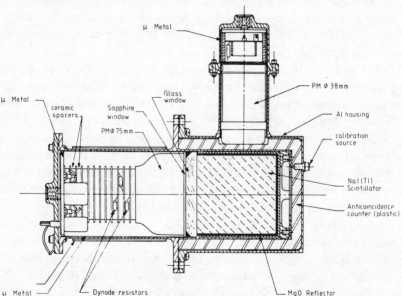

from the γ-rays from natural radionuclides, ^{40}K, ^{232}Th and ^{238}U, γ-ray lines of isotopes produced by the bombardment of the moon's surface by energetic cosmic and solar particles (mainly protons) can be identified. The main elements found by this analysis are O, Si, Fe, Ti, Mg, Al and Ca. The composition of 10% of the moon's surface has been explored in this way. These measurements were supplemented by chemical and mass-spectrometric analyses of samples from the moon's surface brought back by the crews of Apollo moon-landing missions. One unexpected phenomenon discovered by satellite-based γ-ray telescopes were γ-ray bursts in the energy range between 0.2 and 1 MeV, where the intensity of radiation increases by factors of 10 to 100 within time intervals of 0.1 to 100 s. It is not clear yet whether these γ-ray bursts originate within our galaxy or outside, and what is the mechanism generating them.

The exploration of the fluxes of cosmic-ray nuclei at ultrahigh energies is another example of the application of advanced detector technology in space sciences. Such an experiment has been devised by a group at Chicago [SW 82, ME 84]. Fig. 8.6 is a sketch of this cosmic-ray telescope which is to be flown on the Space Shuttle in 1985. It consists of two large hemispherical gas Cherenkov counters separated by several layers of transition radiation

Fig. 8.5. Pulse height spectrum of γ-rays from the surface of the moon, measured by the instrument shown in fig. 8.4 [TR 77]. Full pulse height scale corresponds to about 9 MeV of energy.

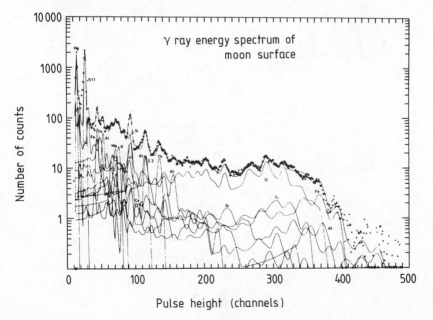

detectors. The aim of the experiment is to measure the energy spectra of nuclei from lithium to iron $(3 \leqslant Z \leqslant 26)$ over an energy range from several hundred GeV/nucleon to several TeV/nucleon, corresponding to Lorentz factors γ of 500 to 5000. The TR radiator consists of mats of thin plastic fibres $(2-6 \, \mu m$ dia.) used commercially for heat isolation ('Thinsulate M400'). The TR yield measured in four proportional chambers filled with $25\% \, Xe + 15\% \, CH_4 + 60\%$ He, each of them behind 20 cm of radiator material, is shown in fig. 8.7 as a function of the Lorentz factor γ. A steep rise of this yield is observed for $\gamma > 500$. Using this TR signal, an energy is assigned to each incoming nucleus. As an additional factor, the gas threshold Cherenkov counters can help to separate nuclei of different mass in the range 100 GeV/nucleon.

8.4 A detector for ion–atom collisions

The probability for ionization of inner atomic shells by collisions of accelerated ions with atoms can be measured by detecting the X-rays from subsequent electronic transitions in the ionized atom. The typical X-ray detectors for these studies are Ge(Li) or Si(Li) semiconductor counters. In order to study the mechanism of such a reaction in detail, additional information about the dependence of the cross-section on the scattering angle θ of the ion and therefore on the impact parameter b is very useful. Such a measurement of the scattering angle is attained, for example, by a

Fig. 8.6. Telescope for measuring the energy spectra of cosmic-ray nuclei [SW 82].

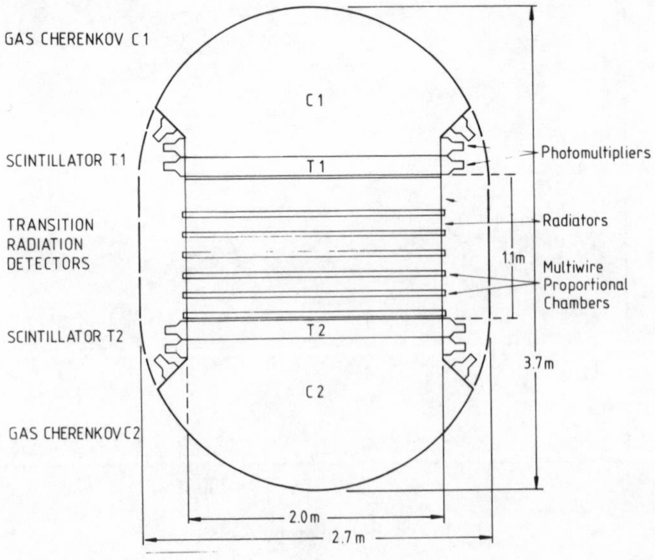

proportional chamber [ST 80]. However, because of the low velocity of the projectile the specific energy loss dE/dx of the heavy ion is so large and its range so short that the chamber has to be operated far below normal pressure. The chamber used is shown in fig. 8.8. The two anode planes each consist of 96 tungsten wires of 20 μm diameter, spaced at 2 mm; the gap between anode and cathode is 5 mm, and the entrance window is made of 2.5 μm Hostaphan material. This chamber was operated for detecting Si, Cu and Ag ions of 20 to 30 MeV energy, with a gas filling of isobutane at 13

Fig. 8.7. Pulse height of transition radiation detectors using foam radiator or fibre radiator as a function of Lorentz factor $\gamma = E/mc^2$ of particles [SW 82].

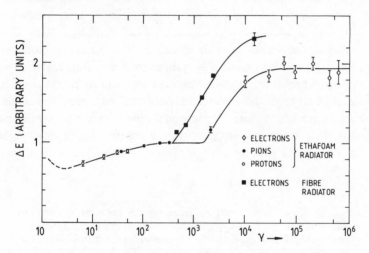

Fig. 8.8. Low-pressure proportional chamber for measuring the scattered ion in ion–atom collisions [ST 80].

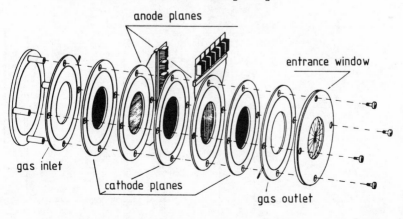

to 22 mbar. The detection efficiency was above 99% and the time resolution 10 ns. In the experiment, the chamber recorded scattered ions in coincidence with X-rays detected in a Si(Li) semiconductor counter, as shown in the geometrical layout in fig. 8.9 [ST 81a]. In this way, the impact parameter dependence of ionization cross-section was measured and information about reaction mechanisms obtained.

8.5 A detector for heavy ion reactions

When the first nuclear physics experiments with heavy ions were done, small semiconductor counters were most widely used as detectors. With increasing energy of the ion projectile, the number of reaction products increases rapidly, in particular if deep inelastic collisions between two nuclei are observed. Initially in such reactions the two heavy nuclei are excited to states far above their ground state. These unstable intermediate states fragment subsequently with emission of neutrons, protons, α-particles or γ-rays or by fission. The number of heavy fragments ranges between two and four; the higher value is obtained in uranium–uranium collisions. In order to identify the reaction products and to reconstruct their kinematics it is possible to detect the secondary particles in coincidence and to measure their ionization energy loss in a counter, their time-of-flight from the target to the counter and their emission angle. For this purpose,

Fig. 8.9. Experimental setup for detecting coincidences between the scattered ion and an X-ray emitted by the atom excited during an ion–atom collision. SBD: surface barrier detector; LPMPWC: low-pressure multiwire proportional chamber. The X-ray is detected in the Si(Li) counter [ST 81a].

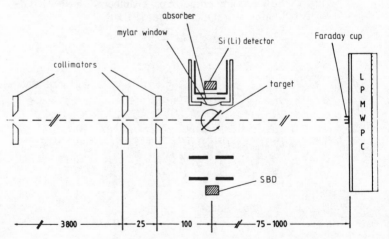

large area ($\sim 1 \, \text{m}^2$) detectors are needed. These can be gas counters developed for this application [e.g. LY 79].

The setup sketched in fig. 8.10 shows the principle of such an experiment used at the Darmstadt heavy ion accelerator UNILAC for the investigation of the reaction $^{40}\text{Ar} + {}^{58}\text{Ni}$ at a laboratory energy of 280 MeV of the incident argon ions. The apparatus includes two γ-ray NaI counters (labelled γ_1 and γ_2) and two detector arms for charged secondary particles. In each arm, the position of the particle is measured by a position sensitive detector (PSD), the time-of-flight by a parallel-plate detector (PPD), and the energy loss by an ionization chamber (DE).

The parallel-plate detectors (see section 4.4) work at low pressure (5–10 Torr) and high electric field strength (5 kV/cm). For heavily ionizing ions a time resolution $\sigma_t \sim 170$ ps is achieved. The position sensitive detectors consist of a pair of ionization chambers with one common cathode plane (see fig. 8.11). At a distance of 90 mm from the cathode, grounded 'Frisch grids' shield the 'θ grid' and the anode planes behind them (section 2.1). Both grid planes are made of wires: for the Frisch grid they run in the x-direction, for the θ grid nearly orthogonal to this direction. A charged particle entering the PSD parallel to the direction produces electron–ion pairs along its track. The electrons drift along the $\pm y$-direction towards the Frisch grid and induce a prompt signal on the cathode. As long as the electron cloud moves between cathode and Frisch grid, the two anode planes are shielded by this grid and do not receive signals. As soon as the

Fig. 8.10. Setup for measuring Ar–Ni inelastic collisions at 280 MeV laboratory energy. γ_1, γ_2: γ-ray detectors; PSD: position sensitive detectors; PPD: parallel-plate detector for time-of-flight measurement; DE: ionisation chamber for energy-loss measurement [LY 79].

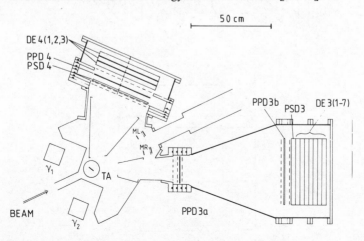

electrons have passed by this grid, they induce a short pulse on the θ grid and a long one on the anode. The time sequence of pulses is shown in fig. 8.11. The x-co-ordinate is measured via the θ grid, by a delay-line readout of its wires. The y-co-ordinate of the track is deduced from the time difference of signals on the cathode and anode, which is proportional to the drift time of electrons in the active volume of the chamber. These co-ordinates are obtained with an accuracy of $\sigma_x \sim 1$ mm.

In the ionization chambers (DE) behind the PSDs, the particles are stopped by ionization energy loss, and this loss is recorded as a charge deposited in the ion chamber.

Fig. 8.11. Position sensitive detector (PSD in fig. 8.10) for detecting reaction products from heavy ion collisions [LY 79].

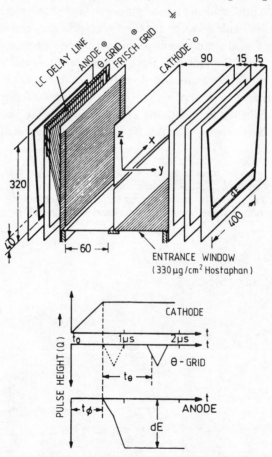

8.6 Detector systems for high-energy experiments

8.6.1 *A detector for hadronic reactions*

In this experiment (NA5 at CERN [NA 5]), a beam of π mesons of 200 to 300 GeV energy impinges on a liquid hydrogen target embedded in the gap of a large dipole magnet (fig. 8.12). Inelastic collisions of this type lead to 15 to 25 secondary hadrons, e.g. π and K mesons, nucleons and heavy mesons. An example for such a reaction is shown in fig. 3.30. These secondary particles are detected in a streamer chamber immediately behind the target. The particle tracks are bent because the streamer chamber is located inside the dipole magnet. The measurement of particle momenta from the track curvature is improved by position measurements in eight large-area spark chambers with magnetostrictive readout. Since the experiment is searching for jets of secondary hadrons at high transverse momentum, a hadron calorimeter [EC 77] downstream of the spark chambers serves for hadron energy measurement and provides a trigger for events with large-energy flux transverse to the beam axis. The angular range of particles accepted by the calorimeter can be varied by moving this instrument back and forth on rails along the beam direction.

8.6.2 *A detector for high-energy neutrinos*

Since the neutrino nucleon total cross-section is only 10^{-36} cm^2 at 100 GeV neutrino energy, neutrino detectors have to be massive. The

Fig. 8.12. Layout (top view) of experiment NA 5 at CERN searching for jets in high-energy hadron–hadron collisions [NA 5].

TOP VIEW

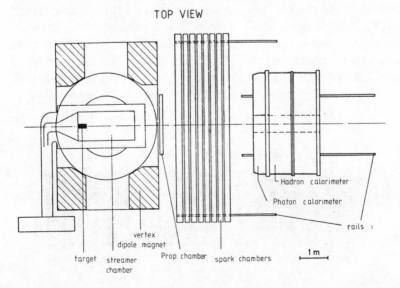

target streamer Prop. chamber spark chambers 1 m
         chamber

vertex
dipole magnet

Hadron calorimeter

Photon calorimeter

rails

detector (fig. 8.13) of the CERN–Dortmund–Heidelberg–Saclay collaboration [HO 78a] uses the target weight of 1500 t of iron, arranged in circular plates of 3.75 m diameter, for three other functions: (i) 75 cm of iron thickness is combined to form a toroidal magnet, (ii) between the iron slabs of 5 cm (for seven magnets) or 15 cm (for eight magnets) eight scintillators viewed by 16 photomultipliers sample the ionization energy deposited by hadronic showers, (iii) particles traversing more than about 2 m of steel are identified as muons, and their momentum is measured from the curvature of their orbit in the toroidal magnetic field. For this purpose large-area hexagonal drift chambers are inserted between the toroids. They have three planes of signal wires strung at 120° azimuthal angle relative to each other [MA 77]. The structure on one drift cell of these chambers is shown in figs 3.8 and 3.9. This detector has registered about 5×10^6 neutrino interactions at the CERN 400 GeV synchrotron. Amongst those were events without muons in the final state (neutral current interactions), with one muon (charged current interactions, quark structure of nucleons) and several muons (production of charmed quarks). For improved measurements of the nucleon structure, a part of the detector was, at a later stage, replaced by magnetized iron toroids with plates of 2.5 cm thickness, whereby the hadron energy resolution was improved. Fig. 8.14 shows a photograph of one such toroid with 3.75 m diameter.

8.6.3   *A detector for electron–positron collisions*

One of the detectors at the PETRA storage rings at DESY (Hamburg) is the JADE detector [BA 79, DR 80] (fig. 8.15). Around the interaction point, tracks are recorded in a central detector (section 7.3) of jet drift chambers (section 2.4) embedded in a 0.45 T solenoidal field. Track coordinates in the (r, φ)-plane are measured with an accuracy of 180 μm, and the momentum resolution is $\Delta P/P^2 = 2.2 \% \, (\text{GeV}/c)^{-1}$. The central detector also measures dE/dx. The field is produced by a 7 cm-thick aluminium coil,

Fig. 8.13. Neutrino detector of the CERN–Dortmund–Heidelberg–Saclay collaboration [HO 78a].

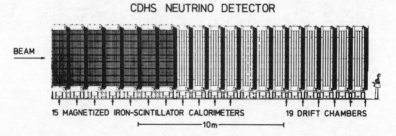

CDHS NEUTRINO DETECTOR

BEAM →

15 MAGNETIZED IRON-SCINTILLATOR CALORIMETERS   19 DRIFT CHAMBERS

|←————————————10m————————————|

3.5 m long, 2 m in diameter. Outside the coil, electron and photon energies are measured in an array of 2520 wedge-shaped lead glass counters, grouped in 30 rings of 84 elements. Together with the lead glass counters on the two endcaps, they cover 90 % of the total solid angle. The magnetic flux of the solenoid returns through the iron covering the cylindrical detector

Fig. 8.14. Magnetized iron–scintillator calorimeter of the CDHS collaboration with 3.75 m diameter iron plates of 2.5 cm thickness, interleaved with planes of scintillator strips.

Fig. 8.15. Section through the JADE experiment detecting electron–positron collisions at centre-of-mass energies between 20 and 40 GeV [BA 79, DR 80].

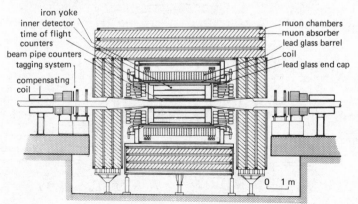

from all sides as a rectangular box. The iron forms part of the muon absorber, of total thickness 785 g/cm^2 or six nuclear absorption lengths. Penetrating tracks are registered by four layers of planar drift chambers.

8.6.4   A detector for proton–antiproton collisions at high energy

The proton–antiproton collider at CERN uses the existing proton synchrotron and devices for cooling and accumulating of antiproton beams in order to produce p$\bar{\text{p}}$ collisions at centre-of-mass energies up to $\sqrt{s}=$ 620 GeV, and tests for operation at $\sqrt{s}=900$ GeV will be done in 1985. One of the detectors at this machine is the UA1 ('Underground Area no. 1') detector. Figs 8.16 and 8.17 give an impression of this enormous detector. The inner detector, with a radius of 1.2 m and a length of 5.7 m, is a cylindrical drift chamber with electron drift path up to 20 cm, embedded in a dipole magnet. The chamber contains about 10 000 signal wires, serving for the measurement of position via drift times and of ionization energy loss dE/dx. Spatial resolution of track co-ordinates amounts to $\sigma_x \sim 250$ μm. The central detector is surrounded by electromagnetic shower counters of the lead–scintillator type, inside the coil of the magnet. The iron magnet yokes are segmented in such a way as to allow scintillation counters to be inserted for hadron calorimetry. For the electromagnetic and the hadronic

Fig. 8.16. Side view of the UA 1 detector observing proton–antiproton collisions at centre-of-mass energies between 540 and 900 GeV.
1: Central detector; 2: large angle barrel hadron calorimeter;
3 and 4: electromagnetic shower counters; 5: end cap hadron calorimeter; 6: muon detector; 7: coil generating dipole field;
8 and 9: small angle detector with chambers and calorimeters;
10: compensator dipole magnets [UA 1].

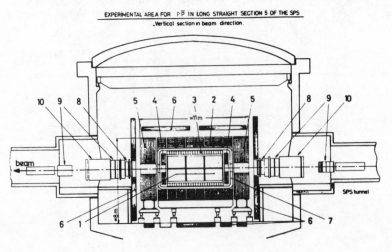

EXPERIMENTAL AREA FOR p$\bar{\text{p}}$ IN LONG STRAIGHT SECTION 5 OF THE SPS
Vertical section in beam direction.

calorimeters, light collection is achieved by wavelength shifter techniques (section 4.3). The outer surface of the iron yokes is covered by two layers of drift tubes, which are able to detect muons traversing the iron.

According to first measurements [AL 82], the particle multiplicity in hadron–hadron collisions seems to rise logarithmically up to 540 GeV

Fig. 8.17. Perspective view of the UA 1 detector with magnet yokes moved apart to show central cylindrical detector.

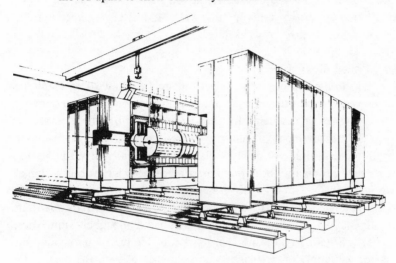

Fig. 8.18. Proton decay detector of the Irvine–Michigan–Brookhaven collaboration (IMB) employing 2048 photomultiplier tubes with 100 mm-diameter photocathodes immersed in a water tank. Cherenkov light cones are used to detect particles with $\beta > 0.75$ [BI 83].

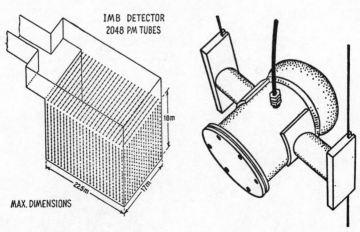

centre-of-mass energy; the typical number of charged secondaries is around 25. Each of the events recorded by this detector contains more than 10^5 bits of information. In spite of this difficulty, the UA1 collaboration, and a second experimental team at this p$\bar{\text{p}}$ machine (UA2), were able in 1983 to isolate from this enormous mass of events about 100 reactions showing the production and decay of the long-sought field quanta of the weak interaction, the vector bosons $W^{\pm}$ and Z^0. The mass of these particles, around 100 proton masses, is exactly the one predicted by the Glashow–Salam–Weinberg theory unifying the weak and electromagnetic interactions into one, the electroweak interaction.

8.7 Proton decay detectors

The unified description of electromagnetic and weak interactions was achieved by the Glashow–Salam–Weinberg model of $SU(2) \times U(1)$ symmetry. Supporting evidence for this theory came from the observation and precise measurement of neutrino-induced neutral-current reactions and from the magnitude of parity violation in the scattering of polarized electrons on protons, and the cornerstone of the theory is the observation of the weak bosons $W^{\pm}$ and Z^0 in p$\bar{\text{p}}$ reactions. Further unification of this electroweak force with the strong quark–quark force described by quantum chromodynamics (QCD) is possible in models based on higher symmetries, like $SU(5)$ or $SO(10)$. In these models, quarks and leptons of one generation belong to one family of particles. A consequence of such theories is the

Fig. 8.19. Detail of calorimeter structure used in the Fréjus proton decay detector [LO 84].

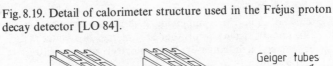
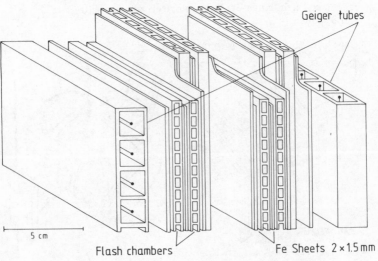

Fig. 8.20. Fréjus proton decay detector during assembly in the cave. The final dimensions are $6 \times 6 \times 9 \text{ m}^3$ and the detector mass is 900 t [ME 85].

decay of protons to mesons and leptons, violating the principle of baryon number conservation. In the $SU(5)$ model, a preferred decay mode is $p \rightarrow e^{+}\pi^{0}$; in other models K mesons emerge as decay products. The expected mean lifetimes for the proton are around 10^{30} years. Large detectors are required in this search for proton decay, and only the cheapest materials, water and steel, are suitable for the purpose.

In water detectors, the Cherenkov radiation of fast particles (electrons and muons of both charges) is detected by a few thousand photomultipliers installed in the water. The Irvine–Michigan–Brookhaven (IMB) detector uses 2408 tubes of 110 mm cathode diameter distributed over the surface of a 6880 t water volume (fig. 8.18). The detector is shielded from cosmic rays by 660 m of rock above the site in the Morton salt mine below Lake Erie. In another experiment, installed in the Kamioka lead mine in Japan, 1000 m below ground, the light yield is increased by employing photomultipliers with spherical, 508 mm diameter, photocathodes (fig. 4.3).

The alternative detection principle is based on a sandwich of iron plates with flash chamber and streamer tube planes. The modular structure of such a detector is shown in fig. 8.19, used for a 800 t detector in the Fréjus tunnel by a French–German collaboration (Saclay–Orsay–Aachen–Wuppertal). This type of detector is sensitive to short charged tracks including slow π or K mesons. Fig. 8.20 is a photograph taken during installation of the Fréjus experiment. The size of the detector will be $6 \times 6 \times 9$ m^{3} after completion.

REFERENCES

AB 81   H. Abramowicz *et al.*, *Nucl. Instr. Meth.* 180 (1981) 429
AD 74   M. Aderholz *et al.*, *Nucl. Instr. Meth.* 118 (1974) 419
AK 77   G. A. Akapdjanov *et al.*, *Nucl. Instr. Meth.* 140 (1977) 441
AL 56   W. P. Allis, *Hdb. d. Physik, Vol. XXI*, Springer Verlag, Heidelberg (1956), p. 383
AL 67   G. D. Alkhazov *et al.*, *Nucl. Instr. Meth.* 48 (1967) 1
AL 69   O. C. Allkofer, *Spark Chambers*, Thiemig Verlag, München (1969)
AL 80   W. W. M. Allison and J. H. Cobb, *Ann. Rev. Nucl. Sci.* 30 (1980) 253
AL 81   W. W. M. Allison, *Phys. Scripta* 23 (1981) 348
AL 82   K. Alpgard *et al.* (UA5 collaboration), *Phys. Lett.* 107 B (1981) 315
AL 83a Aleph–collaboration (Bari–Beijing–CERN–Clermont Ferrand–Copenhagen–
        Demokritos–Dortmund–Ecole–Polytechnique–Edinburgh–Frascati–Glasgow–
        Heidelberg–Imperial College–Lancaster–Marseille–MPI München–Orsay–Pisa–
        Rutherford Laboratory–Saclay–Sheffield–Siegen–Trieste–Westfield College–
        Wisconsin). Technical Report 1983 for CERN–LEP Committee
AL 83b W. W. M. Allison and P. R. S. Wright, Univ. Oxford, Preprint OUNP 35/83
        (1983)
AM 81   U. Amaldi, *Phys. Scripta* 23 (1981) 409
AM 84   S. R. Amendolia *et al.*, *Nucl. Instr. Meth.* 239 A (1985) 192
AN 79   H. Anderhub *et al.*, *Nucl. Instr. Meth.* 166 (1979) 581; 176 (1980) 323
AR 75   X. Artru *et al.*, *Phys. Rev. D* 12 (1975) 1289
AR 81   C. Arnault *et al.*, *Phys. Scripta* 23 (1981) 710
AT 83   W. B. Atwood *et al.*, *Nucl. Instr. Meth.* 206 (1983) 99

BA 56   M. V. Babykin *et al.*, *Sov. J. Atomic Energy* IV (1956) 627
BA 70   G. Bathow *et al.*, *Nucl. Phys. B* 20 (1970) 592
BA 78   B. Barish *et al.*, *IEEE Trans.* NS-25 (1978) 532
BA 79   W. Bartel *et al.*, *Phys. Lett.* 88B (1979) 171
BA 80   M. Barranco Luque *et al.*, *Nucl. Instr. Meth.* 176 (1980) 175
BA 84   A. Bamberger *et al.*, *Nucl. Instr. Meth.* 224 (1984) 408
BA 85   J. Bakken *et al.* (L3 collaboration), submitted to *Nucl. Instr. Meth.* (1985)
BE 30   H. A. Bethe, *Annalen d. Physik* 5 (1930) 325
BE 32   H. A. Bethe, *Z. Physik* 76 (1932) 293
BE 33   H. A. Bethe, *Hdb. Physik* 24 (1933) 518
BE 68   G. Bertolini and A. Coche, *Semiconductor Detectors*, North-Holland,
        Amsterdam (1968)

BE 71   I. B. Berlmann, *Fluorescence Spectra of Aromatic Molecules*, Academic Press, NY and London (1971)
BE 77   *BEBC Users Handbook*, CERN (1977)
BE 81   H. J. Behrend *et al.*, *Phys. Scripta* 23 (1981) 610
BE 83   E. Belau *et al.*, *Nucl. Instr. Meth.* 214 (1983) 253
BI 64   J. B. Birks, *Theory and Practice of Scintillation Counting*, Pergamon Press, Oxford (1964)
BI 75   H. Bichsel and R. P. Saxon, *Phys. Rev. A* 11 (1975) 2186
BI 81   F. Binon *et al.*, *Nucl. Instr. Meth.* 188 (1981) 507
BI 83   R. M. Bionta *et al.*, *Phys. Rev. Lett.* 51 (1983) 27
BL 50   O. Blunck and S. Leisegang, *Z. Physik* 128 (1950) 500
BL 60   J. L. Blankenship and C. J. Borkowski, *IRE Trans.* NS-7, No. 2 (1960) 90
BL 81   W. Blum, private communication (1981)
BL 82   H. Blümer, diploma thesis, Dortmund Univ. (1982)
BO 80a H. Boerner *et al.*, DESY 80/27 (1980)
BO 80b J. Bourotte and B. Sadoulet, *Nucl. Instr. Meth.* 173 (1980) 463
BO 81   H. Bourdinaud and J. C. Thevenin, *Phys. Scripta* 23 (1981) 534
BO 83   A. Bodek *et al.*, *Z. Physik C* 18 (1983) 289
BR 59   S. C. Brown, *Basic Data of Plasma Physics*, MIT Press, Cambridge, MA (1959)
BR 61   W. L. Brown, *IRE Trans.* NS-8, No. 2 (1961)
BR 74   A. Breskin *et al.*, *Nucl. Instr. Meth.* 119 (1974) 9
BR 75   A. Breskin *et al.*, *Nucl. Instr. Meth.* 124 (1975) 189
BR 79   C. Brassard, *Nucl. Instr. Meth.* 162 (1979) 29
BR 81   W. Braunschweig, *Phys. Scripta* 23 (1981) 384
BU 60   R. H. Bube, *Photoconductivity of Solids*, J. Wiley, NY (1960)

CA 74   M. Cantin *et al.*, *Nucl. Instr. Meth.* 118 (1974) 177
CA 81a P. J. Carlson *et al.*, *Phys. Scripta* 23 (1981) 708
CA 81b P. J. Carlson, *Phys. Scripta* 23 (1981) 393
CH 37   P. A. Cherenkov, *Phys. Rev.* 52 (1937) 378
CH 63   G. E. Chikovani *et al.*, *Phys. Lett.* 6 (1963) 254
CH 64   P. A. Cherenkov, I. M. Frank and I. E. Tamm, *Nobel Lectures in Physics*, Elsevier, NY (1964)
CH 68   G. Charpak *et al.*, *Nucl. Instr. Meth.* 62 (1968) 262
CH 70a G. Charpak *et al.*, *Nucl. Instr. Meth.* 80 (1970) 13
CH 70b G. Charpak, *Ann. Rev. Nucl. Sci.* 20 (1970) 195
CH 72   G. Charpak, *Decouverte* (Feb. 1972), p. 9
CH 78a G. Charpak *et al.*, *Nucl. Instr. Meth.* 148 (1978) 471
CH 78b Y. Chan *et al.*, *IEEE Trans.* NS-25 (1978) 333
CO 55   M. Conversi and A. Gozzini, *Nuovo Cim.* 2 (1955) 189
CO 75   J. H. Cobb, PhD thesis, Univ. Oxford (1975)
CO 76   J. H. Cobb *et al.*, *Nucl. Instr. Meth.* 133 (1976) 315
CO 77   J. H. Cobb *et al.*, *Nucl. Instr. Meth.* 140 (1977) 413
CO 78   M. Conversi and L. Federici, *Nucl. Instr. Meth.* 151 (1978) 93
CO 80   V. Commichau *et al.*, *Nucl. Instr. Meth.* 176 (1980) 325
CO 81   D. Cockerill *et al.*, *Phys. Scripta* 23 (1981) 649
CR 60   J. W. Cronin and G. Renninger, *Proc. Int. Conf. on Instrumentation for High Energy Physics*, Berkeley (1960), p. 271
CR 62   D. F. Crawford and H. Messel, *Phys. Rev.* 128 (1962) 352
CU 71   W. Cunitz *et al.*, *Nucl. Instr. Meth.* 91 (1971) 211

DA 79   W. Davies-White *et al.*, *Nucl. Instr. Meth.* 160 (1979) 227
DA 83   M. Danilov *et al.*, *Nucl. Instr. Meth.* 217 (1983) 153

DE 66   G. Dearnaley and D. C. Northrop, *Semiconductor Detectors for Nuclear Radiations*, J. Wiley, NY (1966)
DE 82   M. De Salvo and R. De Salvo, *Nucl. Instr. Meth.* 201 (1982) 357
DH 77   S. Dhawan and R. Majka, *IEEE Trans.* NS-24 (1977) 270
DI 78   M. Dine *et al.*, Fermilab proposal 490 (1978)
DI 79   P. Dishaw, PhD thesis, Stanford Univ. (1979)
DI 80   A. N. Diddens *et al.*, *Nucl. Instr. Meth.* 178 (1980) 27
DR 80   H. Drumm *et al.*, *Nucl. Instr. Meth.* 176 (1980) 333
DY 81   M. Dykes *et al.*, *Nucl. Instr. Meth.* 179 (1981) 487

EC 77   V. Eckhardt *et al.*, *Nucl. Instr. Meth.* 143 (1977) 235
EC 80   V. Eckhardt, private communication (1980)
EI 79   F. R. Eisler, *Nucl. Instr. Meth.* 163 (1979) 105
EI 80   F. Eisele, K. Kleinknecht, D. Pollmann and B. Renk, Internal Report, Univ. Dortmund (1980)
EK 81   T. Ekelöf *et al.*, *Phys. Scripta* 23 (1971) 718
EN 53   W. N. English and G. C. Hanna, *Can. J. Phys.* 31 (1953) 768
EN 74   J. Engler *et al.*, *Nucl. Instr. Meth.* 120 (1974) 157
ER 72   G. A. Erskine, *Nucl. Instr. Meth.* 105 (1972) 565
ER 77   V. C. Ermilova *et al.*, *Nucl. Instr. Meth.* 145 (1977) 555
EV 58   R. D. Evans, *Handbook of Physics*, Springer Verlag, Heidelberg, Vol. 34, p. 218 (1958)

FA 75   C. W. Fabjan and W. Struczinski, *Phys. Lett.* 57 B (1976) 484
FA 77   C. W. Fabjan *et al.*, *Nucl. Instr. Meth.* 141 (1977) 61
FA 79   D. Fancher *et al.*, *Nucl. Instr. Meth.* 161 (1979) 383
FA 80   C. W. Fabjan and H. G. Fischer, *Rep. Progr. Phys.* 43 (1980) 1003
FA 81   C. W. Fabjan *et al.*, *Nucl. Instr. Meth.* 185 (1981) 119; 216 (1983) 105
FI 75   H. G. Fischer *et al.*, *Proc. Int. Meeting on Proportional and Drift Chambers*, Dubna (1975), (JINR) Report D 13-9164
FI 78a   H. G. Fischer, *Nucl. Instr. Meth.* 156 (1978) 81
FI 78b   C. Fichtel, R. Hofstadter, K. Pinkau *et al.*, Proposal for a high-energy gamma-ray telescope on the gamma-ray observatory (Feb. 1978)
FL 81   G. Flügge *et al.*, *Phys. Scripta* 23 (1981) 499
FR 44   O. Frisch, British Atomic Energy Report BT-49 (1944) (unpublished)
FR 55   W. B. Fretter, *Ann. Rev. Nucl. Sci.* 5 (1955) 145
FU 58   H. W. Fulbright, Ionization chambers in nuclear physics, *Handbook of Physics*, Springer Verlag, Heidelberg (1958)

GA 60   R. L. Garwin, *Rev. Sci. Instr.* 31 (1960) 1010
GA 72   R. L. Garwin *et al.*, SLAC-Pub-1133 (1972)
GA 73   G. M. Garibian, *Proc. 5th Int. Conf. on Instrumentation for High Energy Physics*, Frascati (1973), p. 329
GI 46   V. L. Ginzburg and I. M. Frank, *JETP* 16 (1946) 15
GL 52   D. A. Glaser, *Phys. Rev.* 87 (1952) 665; 91 (1953) 496
GL 58   D. A. Glaser, *Hdb. Physik* 45 (1958) 314
GL 63   R. L. Glückstern, *Nucl. Instr. Meth.* 24 (1963) 381
GR 75   R. L. Grasty, *Geophys.* 40 (1975) 503
GR 80   G. Grayer and J. Homer (Rutherford Lab.), private communication (1980)
GU 82   J. C. Guo *et al.*, *Nucl. Instr. Meth.* 204 (1982) 77

HA 73   F. Harris *et al.*, *Nucl. Instr. Meth.* 107 (1973) 413
HA 74   T. M. Harrington *et al.*, *Nucl. Instr. Meth.* 118 (1974) 401

HA 81a   C. K. Hargrove *et al.*, *Phys. Scripta* 23 (1981) 668
HA 81b   G. Harigel *et al.*, *Nucl. Instr. Meth.* 187 (1981) 363
HA 82   G. Harigel, private communication (1982)
HE 78   R. C. Hertzog, Trans. AIME Conference, Houston, Texas (Oct. 1978)
HE 79   R. L. Heath *et al.*, *Nucl. Instr. Meth.* 162 (1979) 431
HE 81   S. Henning and L. Svensson, *Phys. Scripta* 23 (1981) 697
HE 82   A. Hervé *et al.*, preprint CERN–EP/82–28 (1982)
HI 80   H. Hilke, *Nucl. Instr. Meth.* 174 (1980) 145
HI 81   D. Hitlin, *Phys. Scripta* 23 (1981) 634
HO 76   W. Hofmann *et al.*, *Nucl. Instr. Meth.* 135 (1976) 151
HO 78a   M. Holder *et al.*, *Nucl. Instr. Meth.* 148 (1978) 235
HO 78b   M. Holder *et al.*, *Nucl. Instr. Meth.* 151 (1978) 69
HO 79   W. Hofmann *et al.*, *Nucl. Instr. Meth.* 163 (1979) 77
HU 72   E. B. Hughes *et al.*, Stanford Univ. Report No. 627 (1972)
HU 85   G. Hubricht *et al.*, *Nucl. Instr. Meth.* 228 (1985) 327
HY 83   B. Hyams *et al.*, *Nucl. Instr. Meth.* 205 (1983) 99

JA 66   J. D. Jackson, *Classical Electrodynamics*, J. Wiley, NY (1966)

KA 81   V. Kadansky *et al.*, *Phys. Scripta* 23 (1981) 680
KE 48   I. W. Keuffel, *Phys. Rev.* 73 (1948) 531; *Rev. Sci. Instr.* 20 (1949) 202
KE 70   G. Keil, *Nucl. Instr. Meth.* 83 (1970) 145; 87 (1970) 111
KI 10   S. Kinoshito, *Proc. Roy. Soc. A* 83 (1910) 432
KI 79   I. Kirkbridge, *IEEE Trans.* NS-26 (1979) 1535
KI 81   W. Kienzie, private communication (1981)
KL 70   K. Kleinknecht *et al.*, CERN NP Int. Report 70-18 (1970)
KL 81   P. Klasen *et al.*, *Nucl. Instr. Meth.* 185 (1981) 67
KL 82a   K. Kleinknecht, *Phys. Rep.* 84 (1982) 85
KL 82b   F. Klawonn *et al.*, *Nucl. Instr. Meth.* 195 (1982) 483
KN 74   G. Knies and D. Neuffer, *Nucl. Instr. Meth.* 120 (1974) 1
KN 79   G. F. Knoll, *Radiation Detection and Measurement*, J. Wiley, NY (1979)
KO 81   M. Kobayashi *et al.*, *Nucl. Instr. Meth.* 189 (1981) 629
KU 76   D. E. Kuhl *et al.*, *Radiology* 121 (1976) 405
KU 83   H. Kume *et al.*, *Nucl. Instr. Meth.* 205 (1983) 443

LA 44   L. D. Landau, *J. Exp. Phys.* (*USSR*) 8 (1944) 201
LA 83   J. Layter, private communication (1983)
LE 78a   I. Lehraus *et al.*, *Nucl. Instr. Meth.* 153 (1978) 347
LE 78b   B. Leskovar and C. C. Lo, *IEEE Trans.* NS-25 (1978) 582
LE 81a   I. Lehraus *et al.*, *Phys. Scripta* 23 (1981) 727
LE 81b   I. Lehraus *et al.*, preprint CERN/EF 81-14 (1981)
LE 81c   P. Lecomte *et al.*, *Phys. Scripta* 23 (1981) 377
LE 83a   B. Leskovar, *Recent Advances in High-Speed Photon Detectors*, 6th Int. Conf.
          Laser 83–Opto-Electronics, München (1983)
LE 83b   I. Lehraus, *Proc. Wire Chamber Conf.* Vienna (Feb. 1983), CERN/ EF 83–3
LI 73   J. Litt and R. Meunier, *Ann. Rev. Nucl. Sci.* 23 (1973) 1
LO 61   L. B. Loeb, *Basic Processes of Gaseous Electronics*, University of California
          Press, Berkeley (1961)
LO 75   E. Longo and I. Sestili, *Nucl. Instr. Meth.* 128 (1975) 283
LO 81   E. Lorenz, private communication (1981)
LO 83   E. Lohrmann, *Einführung in die Elementarteilchenphysik*, Teubner, Stuttgart
          (1983)
LO 84   C. Longuemare, private communication (1984)

LU 81   T. Ludlam *et al.*, *Nucl. Instr. Meth.* 180 (1981) 413
LU 84   G. Lutz, private communication (1984)
LY 79   U. Lynen *et al.*, *Nucl. Instr. Meth.* 162 (1979) 657

MA 69   H. D. Maccabee and D. G. Papworth, *Phys. Lett.* A30 (1969) 241
MA 77   G. Marel *et al.*, *Nucl. Instr. Meth.* 141 (1977) 43
MA 78   J. N. Marx and D. R. Nygren, *Physics Today* (Oct. 1978) 46
ME 64   A. E. Metzger *et al.*, *Nature* 204 (1964) 766
ME 84   P. Meyer, private communication (1984)
ME 85   H. Meyer, private communication (1985)
MO 80   L. Montanet, *Proc. XXth Int. Conf. on High Energy Physics*, Madison, Wisconsin (July 1980), p. 863
MU 58   R. B. Murray, *Nucl. Instr. Meth.* 2 (1958) 237

NA 1   Frascati–Milano–Pisa–Roma–Torino–Trieste Coll., S. R. Amendiola *et al.*, CERN proposal, SPSC/74–15/P6
NA 5   Bari–Cracow–Liverpool–München (MPI)–Nijmegen–Coll., CERN proposal SPSC/75–1/P37
NA 65   H. H. Nagel, *Z. Physik* 186 (1965) 319
NA 70   R. Nathan and M. Mee, *Phys. Sol.* A2 (1970) 67
NA 72   Y. Nakato *et al.*, *Bull. Chem. Soc. Japan* 45 (1972) 1299
NE 66   H. Neuert, *Kernphysikalische Meßverfahren*, G. Braun, Karlsruhe (1966)
NE 75   O. H. Nestor and C. N. Huang, *IEEE Trans.* NS-22 (1975) 68
NY 74   D. R. Nygren, LBL Int. Report (Feb. 1974)
NY 81   D. R. Nygren, *Phys. Scripta* 23 (1981) 584

PA 68   J. H. Parker and J. J. Lowke, *Phys. Rev.* 181 (1968) 290
PA 75   V. Palladino and B. Sadoulet, *Nucl. Instr. Meth.* 128 (1975) 323
PA 78   Particle Data Group, *Phys. Lett.* 75B (1981) 1
PA 80   R. Partridge *et al.*, *Phys. Rev. Lett.* 44 (1980) 712
PE 76   Proposal for a PEP facility based on the time projection chamber (TPC), Johns Hopkins Univ.; Lawrence Berkeley Lab.; Univ. of Calif., Los Angeles; Univ. of Calif., Riverside; Yale Univ; PEP Exp. No. 4, SLAC Pub–5012 (1976)
PE 82   PEP–4 collaboration, Proposal to modify the time projection chamber in order to eliminate track distortions, Report TPC–LBL–82–84, Lawrence Berkeley Lab. (Sept. 1984)
PH 77   M. Phelps, *Sem. Nucl. Med.* 7, No. 4 (1977) 337
PH 78   *Philips Data Handbook, Part 9* (1978)
PR 58   W. Price, *Nuclear Radiation Detection*, McGraw-Hill, NY (1958)
PR 80   Y. D. Prokoshkin, *Proc. of Second ICFA Workshop*, Les Diablerets (Oct. 1979); CERN Report (June 1980)

RA 74   V. Radeky, *IEEE Trans.* NS-21 (1974) 51
RA 83   C. Raine *et al.*, *Nucl. Instr. Meth.* 217 (1983) 305
RE 11   R. Reiganum, *Z. Physik* 12 (1911) 1076
RI 54   M. Rich and R. Madey, *Range Energy Tables*, UCRL Report No. 2301 (1954)
RI 74   P. Rice-Evans, *Spark, Streamer, Proportional and Drift Chambers*, London (1974)
RO 52   B. Rossi, *High Energy Particles*, Prentice Hall, NY (1952)
RO 56   H. H. Rossi and G. Failla, *Nucleonics* 14, No. 2 (1965) 32
RO 72   R. E. Robson, *Austr. J. Phys.* 25 (1972) 625
RU 64   J. G. Rutherglen, *Progr. Nucl. Phys.* 9 (1964) 3

SA 64   A. Sayres and M. Coppola, *Rev. Sci. Instr.* 35 (1964) 431

SA 77   F. Sauli, *Principles of Operation of Multiwire, Proportional and Drift Chambers*, CERN Report 77–09 (1977)

SA 80   J. Sandweiss, XXth Int. Conf. on High Energy Physics, Madison, Wisconsin (July 1980)

SA 81   F. Sauli, *Phys. Scripta* 23 (1981) 526

SC 71   P. Schilly *et al.*, *Nucl. Instr. Meth.* 91 (1971) 221

SC 78   G. Schultz and J. Gresser, *Nucl. Instr. Meth.* 151 (1978) 413

SC 79   L. S. Schröder, *Nucl. Instr. Meth.* 162 (1979) 395

SC 80   B. Schmidt, Diploma thesis, Univ. Heidelberg (1980)

SC 82   M. A. Schneegans *et al.*, *Nucl. Instr. Meth.* 139 (1982) 445

SC 84   W. Schmidt-Parzefall, private communication (1984)

SE 77   J. Seguinot and T. Ypsilantis, *Nucl. Instr. Meth.* 142 (1977) 377

SE 79   W. Selove *et al.*, *Nucl. Instr. Meth.* 161 (1979) 233

SH 50   W. Shockley, *Electrons and Holes in Semiconductors*, van Nostrand, NY (1950)

SH 51   W. A. Shurcliff, *J. Opt. Soc. Am.* 41 (1951) 209

SH 75   E. Shibamura *et al.*, *Nucl. Instr. Meth.* 131 (1975) 249

ST 52   R. M. Sternheimer, *Phys. Rev.* 88 (1952) 851

ST 53   H. Staub, in *Experimental Nuclear Physics* (E. Segré, ed.), J. Wiley, NY (1953), Vol. I, p. 1

ST 71   R. M. Sternheimer and R. F. Peierls, *Phys. Rev.* B3 (1971) 3681

ST 80   J. Stähler and G. Presser, *Nucl. Instr. Meth.* 177 (1980) 427

ST 81   J. Stähler and G. Presser, *Nucl. Instr. Meth.* 189 (1981) 603

ST 81*b* S. Stone, *Phys. Scripta* 23 (1981) 605

SW 82   S. P. Swordy, *Nucl. Instr. Meth.* 193 (1982) 591

SZ 81   S. M. Sze, *Physics of Semiconductor Devices*, J. Wiley, 2nd edition, NY (1981)

TA 78   F. E. Taylor *et al.*, *IEEE Trans.* NS-25 (1978) 312

TI 83   R. Tiemann, Diploma thesis, Univ. Dortmund (1983)

TR 69   T. Trippe, CERN NP Int. Report 69–18 (1969)

TR 77   J. I. Trombka *et al.*, *Astrophys. J.* 212 (1977) 925

UA 1   Aachen–Annecy–Birmingham–CERN–London–Paris–Riverside–Rutherford–Saclay–Vienna collaboration, CERN proposal SPSC/78–6, SPSC/P92 (1978)

UA 5   Bonn–Brussels–Cambridge–CERN–Stockholm collaboration, CERN proposal SPSC/78–70/P108 (1978)

VA 70   *Valvo Photomultiplier Buch*, Hamburg (April 1970)

WA 21   Birmingham–Bonn–CERN–London–Munich–Oxford collaboration using BEBC at CERN, Exp. WA 21 (1979)

WA 67   J. H. Ward and B. J. Thompson, *J. Opt. Soc. Am.* 57 (1967) 275

WA 71   A. H. Walenta *et al.*, *Nucl. Instr. Meth.* 92 (1971) 373

WA 79   A. H. Walenta *et al.*, *Nucl. Instr. Meth.* 161 (1979) 45

WA 81*a* A. H. Walenta, *Phys. Scripta* 23 (1981) 354

WA 81*b* A. Wagner, *Phys. Scripta* 23 (1981) 446

WA 82   A. Wagner, private communication (1982)

WE 66   W. T. Welford, *Appl. Opt.* 5 (1977) 872

WE 81*a* D. Wegener, private communication (1981)

WE 81*b* H. Wenninger, private communication (1981)

WI 57   R. L. Williams, *Can. J. Phys.* 35 (1957) 134

WI 74   W. J. Willis and V. Radeka, *Nucl. Instr. Meth.* 120 (1974) 221

YP 81   T. Ypsilantis, *Phys. Scripta* 23 (1981) 371

INDEX

....